THE ONGOING CHALLENGE OF MANAGING CARBON MONOXIDE POLLUTION IN FAIRBANKS, ALASKA

Interim Report

Committee on Carbon Monoxide Episodes in
Meteorological and Topographical Problem Areas

Board on Environmental Studies and Toxicology

Board on Atmospheric Sciences and Climate

Division on Earth and Life Studies

Transportation Research Board

National Research Council

NATIONAL ACADEMY PRESS
Washington, D.C.

NATIONAL ACADEMY PRESS   2101 Constitution Ave., N.W.   Washington, D.C. 20418

NOTICE: The project that is the subject of this report was approved by the Governing Board of the National Research Council, whose members are drawn from the councils of the National Academy of Sciences, the National Academy of Engineering, and the Institute of Medicine. The members of the committee responsible for the report were chosen for their special competences and with regard for appropriate balance.

This project was supported by Cooperative Agreement X 82880601-0, between the National Academy of Sciences and the U.S. Environmental Protection Agency. Any opinions, findings, conclusions, or recommendations expressed in this publication are those of the author(s) and do not necessarily reflect the view of the organizations or agencies that provided support for this project.

International Standard Book Number: 0-309-08484-9

Additional copies of this report are available from:

National Academy Press
2101 Constitution Ave., NW
Box 285
Washington, DC 20055

800-624-6242
202-334-3313 (in the Washington metropolitan area)
http://www.nap.edu

Copyright 2002 by the National Academy of Sciences. All rights reserved.

Printed in the United States of America

THE NATIONAL ACADEMIES

National Academy of Sciences
National Academy of Engineering
Institute of Medicine
National Research Council

The **National Academy of Sciences** is a private, nonprofit, self-perpetuating society of distinguished scholars engaged in scientific and engineering research, dedicated to the furtherance of science and technology and to their use for the general welfare. Upon the authority of the charter granted to it by the Congress in 1863, the Academy has a mandate that requires it to advise the federal government on scientific and technical matters. Dr. Bruce M. Alberts is president of the National Academy of Sciences.

The **National Academy of Engineering** was established in 1964, under the charter of the National Academy of Sciences, as a parallel organization of outstanding engineers. It is autonomous in its administration and in the selection of its members, sharing with the National Academy of Sciences the responsibility for advising the federal government. The National Academy of Engineering also sponsors engineering programs aimed at meeting national needs, encourages education and research, and recognizes the superior achievements of engineers. Dr. Wm. A. Wulf is president of the National Academy of Engineering.

The **Institute of Medicine** was established in 1970 by the National Academy of Sciences to secure the services of eminent members of appropriate professions in the examination of policy matters pertaining to the health of the public. The Institute acts under the responsibility given to the National Academy of Sciences by its congressional charter to be an adviser to the federal government and, upon its own initiative, to identify issues of medical care, research, and education. Dr. Harvey V. Fineberg is president of the Institute of Medicine.

The **National Research Council** was organized by the National Academy of Sciences in 1916 to associate the broad community of science and technology with the Academy's purposes of furthering knowledge and advising the federal government. Functioning in accordance with general policies determined by the Academy, the Council has become the principal operating agency of both the National Academy of Sciences and the National Academy of Engineering in providing services to the government, the public, and the scientific and engineering communities. The Council is administered jointly by both Academies and the Institute of Medicine. Dr. Bruce M. Alberts and Dr. Wm. A. Wulf are chairman and vice chairman, respectively, of the National Research Council.

COMMITTEE ON CARBON MONOXIDE EPISODES IN METEOROLOGICAL AND TOPOGRAPHICAL PROBLEM AREAS

Members

ARMISTEAD G. RUSSELL *(Chair)*, Georgia Institute of Technology, Atlanta
ROGER ATKINSON, University of California, Riverside
SUE ANN BOWLING, University of Alaska (Retired), Fairbanks
STEVEN D. COLOME, University of California, Los Angeles
NAIHUA DUAN, University of California, Los Angeles
GERALD GALLAGHER, J Gallagher and Associates, Inc., Englewood, Colorado
RANDALL L. GUENSLER, Georgia Institute of Technology, Atlanta
SUSAN L. HANDY, University of Texas, Austin
SIMONE HOCHGREB, Exponent, Natick, Massachusetts
SANDRA N. MOHR, Consultant, Gillette, New Jersey
ROGER A. PIELKE SR., Colorado State University, Fort Collins
KARL J. SPRINGER, Southwest Research Institute (Retired), San Antonio, Texas
ROGER WAYSON, University of Central Florida, Orlando, Florida

Project Staff

K. JOHN HOLMES, Senior Staff Officer
RAYMOND WASSEL, Senior Program Director
NANCY HUMPHREY, Senior Staff Officer
CHAD TOLMAN, Staff Officer
LAURIE GELLER, Staff Officer
AMANDA STAUDT, Postdoctoral Research Associate
NORMAN GROSSBLATT, Editor
KELLY CLARK, Editorial Assistant
RAMYA CHARI, Project Assistant

Sponsor

U.S. ENVIRONMENTAL PROTECTION AGENCY

BOARD ON ENVIRONMENTAL STUDIES AND TOXICOLOGY

Members

GORDON ORIANS *(Chair)*, University of Washington, Seattle
JOHN DOULL *(Vice Chair)*, University of Kansas Medical Center, Kansas City
DAVID ALLEN, University of Texas, Austin
INGRID C. BURKE, Colorado State University, Fort Collins
THOMAS BURKE, Johns Hopkins University, Baltimore, Maryland
WILLIAM L. CHAMEIDES, Georgia Institute of Technology, Atlanta
CHRISTOPHER B. FIELD, Carnegie Institute of Washington, Stanford, California
J. PAUL GILMAN, Celera Genomics, Rockville, Maryland
DANIEL S. GREENBAUM, Health Effects Institute, Cambridge, Massachusetts
BRUCE D. HAMMOCK, University of California, Davis
ROGENE HENDERSON, Lovelace Respiratory Research Institute, Albuquerque, New Mexico
CAROL HENRY, American Chemistry Council, Arlington, Virginia
ROBERT HUGGETT, Michigan State University, East Lansing
JAMES H. JOHNSON, Howard University, Washington, D.C.
JAMES F. KITCHELL, University of Wisconsin, Madison
DANIEL KREWSKI, University of Ottawa, Ottawa, Ontario
JAMES A. MACMAHON, Utah State University, Logan
WILLEM F. PASSCHIER, Health Council of The Netherlands, The Hague
ANN POWERS, Pace University School of Law, White Plains, New York
LOUISE M. RYAN, Harvard University, Boston, Massachusetts
KIRK SMITH, University of California, Berkeley
LISA SPEER, Natural Resources Defense Council, New York

Senior Staff

JAMES J. REISA, Director
DAVID J. POLICANSKY, Associate Director and Senior Program Director for Applied Ecology
RAYMOND A. WASSEL, Senior Program Director for Environmental Sciences and Engineering
KULBIR BAKSHI, Program Director for the Committee on Toxicology
ROBERTA M. WEDGE, Program Director for Risk Analysis
K. JOHN HOLMES, Senior Staff Officer
SUSAN N.J. MARTEL, Senior Staff Officer
SUZANNE VAN DRUNICK, Senior Staff Officer
RUTH E. CROSSGROVE, Managing Editor

BOARD ON ATMOSPHERIC SCIENCES AND CLIMATE

Members

ERIC J. BARRON (*Chair*), Pennsylvania State University, University Park
SUSAN K. AVERY, University of Colorado, Boulder
RAYMOND J. BAN, The Weather Channel, Inc., Atlanta, Georgia
HOWARD B. BLUESTEIN, University of Oklahoma, Norman
STEVEN F. CLIFFORD, National Oceanic and Atmospheric Administration, Boulder, Colorado
GEORGE L. FREDERICK, Vaisala Meteorological Systems, Inc., Boulder, Colorado
JUDITH L. LEAN, Naval Research Laboratory, Washington, DC
MARGARET A. LEMONE, National Center for Atmospheric Research, Boulder, Colorado
MARIO J. MOLINA, Massachusetts Institute of Technology, Cambridge
ROGER A. PIELKE, JR., University of Colorado, Boulder
MICHAEL J. PRATHER, University of California, Irvine
WILLIAM J. RANDEL, National Center for Atmospheric Research, Boulder, Colorado
ROBERT T. RYAN, WRC-TV, Washington, DC
THOMAS F. TASCIONE, Sterling Software, Inc., Bellevue, Nebraska
ROBERT A. WELLER, Woods Hole Oceanographic Institution, Woods Hole, Massachusetts
ERIC F. WOOD, Princeton University, Princeton, New Jersey

TRANSPORTATION RESEARCH BOARD 2000 EXECUTIVE COMMITTEE

Members

JOHN M. SAMUELS (*Chair*), Norfolk Southern Corporation, Norfolk, Virginia
THOMAS R. WARNE (*Vice Chair*), Utah Department of Transportation, Salt Lake City
ROBERT E. SKINNER, JR. (Executive Director), National Research Council, Washington, D.C.
WILLIAM D. ANKNER, Rhode Island Dept. of Transportation, Providence
THOMAS F. BARRY, JR., Florida Department of Transportation, Tallahassee
JACK E. BUFFINGTON, University of Arkansas, Fayetteville
SARAH C. CAMPBELL, TransManagement, Inc., Washington, D.C.
E. DEAN CARLSON, Kansas Department of Transportation, Topeka
JOANNE CASEY, Intermodal Association of North America, Greenbelt, Maryland
JAMES C. CODELL III, Kentucky Transportation Cabinet, Frankfort
JOHN L. CRAIG, Nebraska Department of Roads, Lincoln
ROBERT A. FROSCH, Harvard University, Cambridge, Massachusetts
GORMAN GILBERT, Oklahoma State University, Stillwater
GENEVIEVE GIULIANO, University of Southern California, Los Angeles
LESTER A. HOEL, University of Virginia, Charlottesville
H. THOMAS KORNEGAY, Port of Houston Authority, Houston, Texas
BRADLEY L. MALLORY, Pennsylvania Department of Transportation, Harrisburg
MICHAEL D. MEYER, Georgia Institute of Technology, Atlanta
JEFF P. MORALES, California Department of Transportation, Sacramento
JEFFREY R. MORELAND, Burlington Northern Santa Fe Railway, Fort Worth, Texas
JOHN P. POORMAN, Capital District Transportation Committee, Albany, New York
CATHERINE L. ROSS, Georgia Regional Transportation Agency, Atlanta
WAYNE SHACKELFORD, Gresham Smith & Partners, Alpharetta, Georgia
PAUL P. SKOUTELAS, Port Authority of Allegheny County, Pittsburgh, Pennsylvania
MICHAEL S. TOWNES, Transportation District Commission of Hampton Roads, Hampton, Virginia
MARTIN WACHS, University of California, Berkeley
MICHAEL W. WICKHAM, Roadway Express, Inc., Akron, Ohio
JAMES A. WILDING, Metropolitan Washington Airports Authority, Washington, D.C.
M. GORDON WOLMAN, The Johns Hopkins University, Baltimore, Maryland

OTHER REPORTS OF THE
BOARD ON ENVIRONMENTAL STUDIES AND TOXICOLOGY

The Airliner Cabin Environment and Health of Passengers and Crew (2002)
Arsenic in Drinking Water: 2001 Update (2001)
Evaluating Vehicle Emissions Inspection and Maintenance Programs (2001)
Compensating for Wetland Losses Under the Clean Water Act (2001)
A Risk-Management Strategy for PCB-Contaminated Sediments (2001)
Toxicological Effects of Methylmercury (2000)
Strengthening Science at the U.S. Environmental Protection Agency: Research-Management and Peer-Review Practices (2000)
Scientific Frontiers in Developmental Toxicology and Risk Assessment (2000)
Copper in Drinking Water (2000)
Ecological Indicators for the Nation (2000)
Waste Incineration and Public Health (1999)
Hormonally Active Agents in the Environment (1999)
Research Priorities for Airborne Particulate Matter (3 reports, 1998-2001)
Ozone-Forming Potential of Reformulated Gasoline (1999)
Arsenic in Drinking Water (1999)
Brucellosis in the Greater Yellowstone Area (1998)
The National Research Council's Committee on Toxicology: The First 50 Years (1997)
Carcinogens and Anticarcinogens in the Human Diet (1996)
Upstream: Salmon and Society in the Pacific Northwest (1996)
Science and the Endangered Species Act (1995)
Wetlands: Characteristics and Boundaries (1995)
Biologic Markers (5 reports, 1989-1995)
Review of EPA's Environmental Monitoring and Assessment Program (3 reports, 1994-1995)
Science and Judgment in Risk Assessment (1994)
Pesticides in the Diets of Infants and Children (1993)
Protecting Visibility in National Parks and Wilderness Areas (1993)
Dolphins and the Tuna Industry (1992)
Science and the National Parks (1992)
Assessment of the U.S. Outer Continental Shelf Environmental Studies Program, Volumes I-IV (1991-1993)
Human Exposure Assessment for Airborne Pollutants (1991)
Rethinking the Ozone Problem in Urban and Regional Air Pollution (1991)
Decline of the Sea Turtles (1990)

Copies of these reports may be ordered from the National Academy Press
(800) 624-6242 or (202) 334-3313
www.nap.edu

Acknowledgment of Review Participants

This report has been reviewed in draft form by individuals chosen for their diverse perspectives and technical expertise, in accordance with procedures approved by the NRC's Report Review Committee. The purpose of this independent review is to provide candid and critical comments that will assist the institution in making its published report as sound as possible and to ensure that the report meets institutional standards for objectivity, evidence, and responsiveness to the study charge. The review comments and draft manuscript remain confidential to protect the integrity of the deliberative process. We wish to thank the following individuals for their review of this report:

John C. Bailar III, University of Chicago
Lenora Bohren, Colorado State University
Gregory J. Dana, Alliance of Automobile Manufacturers
Robert Dulla, Sierra Research, Inc.
Robert Gibbons, University of Illinois, Chicago
Judith A. Graham, American Chemistry Council
Arthur Hussey, Northern Alaska Environmental Center
Robert F. Klausmeier, de la Torre Klausmeier Consulting, Inc.
Paul W. Ludden, University of Wisconsin, Madison
Jeffery S. Tilley, University of Alaska, Fairbanks

Although the reviewers listed above have provided many constructive comments and suggestions, they were not asked to endorse the conclusions or recommendations, nor did they see the final draft of the report before its release. The review of this report was overseen by F. Sherwood Rowland, University of California, Irvine. Appointed by the NRC, he was responsible for making certain that an independent examination of this report was carried out in accordance with institutional procedures and that all review comments were carefully considered. Responsibility for the final content of this report rests entirely with the authoring committee and the institution.

Preface

Carbon monoxide (CO) is a toxic air pollutant produced largely from vehicle emissions. Breathing CO at high concentrations leads to reduced oxygen transport by hemoglobin, which has health effects that include impaired reaction timing, headaches, lightheadedness, nausea, vomiting, weakness, clouding of consciousness, coma, and, at high enough concentrations and long enough exposure, death. In recognition of those health effects, the U.S. Environmental Protection Agency (EPA), as directed by the Clean Air Act, established the health-based National Ambient Air Quality Standards (NAAQS) for CO in 1971.

Most areas that were previously designated as "nonattainment" areas have come into compliance with the NAAQS for CO, but some locations still have difficulty in attaining the CO standards. Those locations tend to have topographical or meteorological characteristics that exacerbate pollution. In view of the challenges posed for some areas to attain compliance with the NAAQS for CO, congress asked the National Research Council to investigate the problem of CO in areas with meteorological and topographical problems. This interim report deals specifically with Fairbanks, Alaska. Fairbanks was chosen as a case study because its meteorological and topographical characteristics make it susceptible to severe winter inversions that trap CO and other pollutants at ground level.

In preparing this report, the committee gathered information and conducted analyses of existing data but did not make any new measurements. The committee met in Fairbanks in August 2001 and again in December 2001, when they were able to observe first-hand the meteorological conditions and emissions patterns that could cause a CO episode. During their second trip to Fairbanks, a public hearing was held to assist the committee in issue identification. Members of the community were invited to share their perspectives and concerns with the committee at that time. The committee would like to acknowledge those who took the time to participate in that hearing.

Many people assisted the committee by providing information related to issues addressed in this report. I gratefully acknowledge William Boycott, Williams Alaska Petroleum Inc.; John Cabaniss, Association of International Automobile Manufacturers; Gregory Dana, Alliance of Automobile Manufacturers; Robert Dulla, Sierra Research; Laurence Elmore, EPA; Mary Ellen Gordian, University of Alaska, Anchorage; Gerald Guay, Alaska Department of Environmental Conservation; Nadine Hargesheimer, Fairbanks North Star Borough; Gregory Henderson, Tesoro Alaska; Ronald King, Alaska Department of Environmental Conservation; Max Lyon, Fairbanks North Star Borough; John Middaugh, Alaska Department of Health and Public Services; Steven Morris, municipality of Anchorage; William Neff, National Oceanic and Atmospheric Administration; Paul Prusak, Alaska Department of Transportation and Public Facilities; Paul Rossow, Fairbanks North Star Borough; Leonard Verrelli, state of Alaska; and Jimmie Williams, Ceramics, Environmental Technologies Development, Corning, Inc.; Aaron Owens, DuPont Central Research and Development, Wilmington, Delaware; and Susan Alber, University of California, Los Angeles.

I am also grateful for the assistance of the National Research Council staff in the preparation of this report. The committee was ably assisted by K. John Holmes in his role as project director. The committee also acknowledges Raymond A. Wassel, senior program director for environmental sciences and engineering in the Board on Environmental Studies and Toxicology (BEST). We thank the other staff members who contributed to this report, including Warren Muir, executive director of the Division on Earth and Life Studies; James J. Reisa, director of BEST; Nancy Humphrey, senior staff officer with the Transportation Research Board; Chad Tolman, staff officer with BEST; Amanda Staudt, post-doctoral research associate with BEST; Norman Grossblatt, editor; Mirsada Karalic-Loncarevic, information specialist with BEST; and Ramya Chari, Jennifer Saunders, and Emily Smail, project assistants with BEST.

Finally, I would like to thank all the members of the committee for their expertise and dedicated effort throughout the study.

> Armistead Russell, Ph.D.
> *Chair*, Committee on Carbon
> Monoxide Episodes in Meteorological
> and Topographical Problem Areas

Contents

Summary..1

1   Sources and Effects of Carbon Monoxide Emissions..........................19
    The Problem, 19
    Charge to the Committee, 20
    Health Effects of CO, 22
    National Air Quality Status for Ambient CO, 28
    Meteorology and Topography in CO Nonattainment Areas, 31
    National Inventory of CO Emissions, 37
    CO Emissions from Vehicles, 38
    Air Quality Models, 44
    Summary, 47

2   Fairbanks Case Study..49
    Geographical, Meteorological, and Societal Context, 50
    CO Data, 57
    Emissions and Vehicle Characteristics, 72
    CO Control Programs, 77
    Simple Box Model of the Borough, 98
    Summary, 107

3   Implications of the Fairbanks Case Study.......................................109
    Prospects for Continued Attainment, 109

Relationship of Fairbanks Case Study to Other Nonattainment Areas, 111

References 113

Glossary 122

Appendix. Biographical Information on the Committee on Carbon Monoxide Episodes in Meteorological and Topographical Problem Areas 131

The Ongoing Challenge of Managing Carbon Monoxide Pollution in Fairbanks, Alaska

Summary

Carbon monoxide (CO) is a colorless, odorless gas that can produce serious adverse health effects. Exposure to high concentrations can be fatal. At lower concentrations that can occur in the ambient environment, the effects of CO include increased risk of chest pain and hospitalization for persons with coronary artery disease. Recognizing the public-health hazards of elevated ambient concentrations, the U.S. Environmental Protection Agency (EPA) set National Ambient Air Quality Standards (NAAQS) for CO in 1971.[1] When the concentration of CO in ambient air exceeds health standards, it is due mainly to incomplete combustion of gasoline by light-duty vehicles, such as passenger cars and pickup trucks. EPA accordingly has set increasingly stringent standards for CO emissions from vehicles and has mandated vehicle emissions inspection and maintenance (I/M) programs and the use of oxygenated fuels to reduce emissions in areas with ambient-CO problems.

The controls are working. In the 1970s, numerous cities exceeded the CO health standards. Now, only a handful of areas fail to meet the NAAQS for

---

[1] The standards for ambient air concentrations of CO were set at 9 parts per million (ppm) for an 8-hour (h) average and 35 ppm for a 1-h average. These standards were set to protect public health with "an adequate margin of safety," as specified in the Clean Air Act. A violation of an NAAQS occurs on the second and all later exceedances of the standard in a calendar year. Only the 8-h standard of 9 ppm is currently exceeded in a few locations in the country.

CO,[2] and they are experiencing fewer violations as automotive controls continue to reduce emissions. However, a few areas, including Lynwood and Calexico, California, and Fairbanks, Alaska, have continued to experience violations over the last five years. Those three areas have very different physical and demographic characteristics, and the reasons for their continuing problems differ greatly.

Fairbanks is an extreme example of the roles that meteorology and land topography play in producing air quality problems. In winter, Fairbanks is subject to extreme atmosphere inversions, at times recording inversion strengths of as much as 30°C (86°F) per 100 m of altitude.[3] In addition, Fairbanks is situated in a three-sided bowl, surrounded by the Yukon-Tanana uplands; the bowl opens to the Tanana River Flats toward the south and southeast. Although Fairbanks is not heavily populated and has no major air-pollution-producing industries, its meteorological and topographical characteristics make the city susceptible to high ambient CO concentrations in winter. The atmospheric inversions and low windspeeds that commonly occur during winter are extremely effective in trapping the products of incomplete combustion, including CO, that are emitted at ground level.

The Fairbanks North Star Borough, Alaska Department of Environmental Conservation (ADEC), and EPA have been concerned about the inability of Fairbanks to attain the CO health standard. In the fiscal year 2001 appropriations for EPA, Congress called for the National Research Council (NRC) to conduct an independent study of CO episodes in meteorological and topographical problem areas. The study is to address various potential approaches to predicting, assessing, and managing episodes of high concentrations of CO in such areas. The complete study charge is contained in Chapter 1. Congress also requested that the Fairbanks area be the subject of a case study in an interim report. In response, the NRC formed the Committee on Carbon Mon-

---

[2]Areas that violate an NAAQS for a given pollutant are designated as being in nonattainment for that pollutant. Their nonattainment status triggers many regulatory requirements, including the submission to EPA of an attainment plan, known as a state implementation plan (SIP), which describes the strategies an area will use to come into compliance.

[3]Inversions occur when the temperature of the atmosphere increases with height. Combined with low windspeeds, this prevents air circulation because colder air is trapped near the ground by the warmer air above. A temperature increase of several degrees per 100 m is considered a strong inversion.

oxide Episodes in Meteorological and Topographical Problem Areas, which has prepared this interim report. The final report will more broadly address approaches to predicting, assessing, and managing CO episodes in other problem areas in addition to Fairbanks.

THE COMMITTEE'S APPROACH TO ITS CHARGE

In this interim report on Fairbanks, the committee addresses meteorological and topographical conditions that foster pollution episodes, wintertime CO emissions from mobile sources, air quality management options, and statistical methods for tracking progress. The committee also examined monitoring data for ambient CO, including episodes when 8-hour (h) average concentrations exceeded the NAAQS. Although Fairbanks has monitored CO in three downtown locations since 1972 and has a laboratory to study vehicle emissions under cold-weather conditions, data and modeling to assess the spatial and temporal extent of high-CO events are limited for the city. This reduced the ability of the committee to assess the exposure of residents during high-CO episodes. In addition, little exposure or epidemiological data specific to Fairbanks were available to assess the public health impact of those high ambient CO episodes. In the absence of such data, the committee reviewed relevant clinical and epidemiological studies presented in the scientific literature and considered by EPA in assessing the health impact of exposure to ambient CO at concentrations and durations exceeding the NAAQS. In addition, the technical feasibility and potential for emissions reductions of a number of air quality management options are discussed in the body of this report. Promising options available for implementation at the state and local levels are presented in this summary. The committee's recommendations follow and can be found in the body of the report adjacent to further supporting evidence.

FINDINGS AND RECOMMENDATIONS

CO Concentration Trends in Fairbanks

Fairbanks has made great progress in reducing its violations of the 8-h CO health standard. The Fairbanks North Star Borough has worked effectively to reduce CO emissions. It has also benefited greatly from stringent federal vehicle-emissions standards for CO, and the borough and Alaska have invested more resources in studying and characterizing the air quality problem in Fairbanks than other cities of this size in the United States. The number of days annually with violations has been reduced from over 130 during 1973 and 1974 to zero over the last 2 years;[4] that demonstrates the effectiveness of new-vehicle emissions standards, including requirements for certifying vehicles at a low temperature (20°F [-6.7°C]), and of local controls to reduce motor-vehicle emissions.

Despite those efforts, Fairbanks will continue to be susceptible to violating the 8-h CO health standard on some occasions for many years to come because of its unfavorable meteorological and topographical conditions. Those adverse natural conditions might be compounded by future increases in the population of Fairbanks brought about by large pipeline or military construction projects. In such cases, emissions controls may have to be enhanced to offset increases.

Borough officials have argued that Fairbanks should be granted an exemption from the Clean Air Act with regard to the ambient CO health standards because of its extreme meteorological and topographical conditions. However, a similar argument could be made for other regions with regard to a variety of air pollutants. Furthermore, the ambient concentrations of CO observed in Fairbanks have exceeded the level that EPA identified in the health-based standards for the protection of the general population and susceptible individuals.

Health and Exposure

The committee finds that EPA has presented strong scientific evidence in support of the NAAQS describing the health effects of CO on the general population and susceptible individuals. Although ambient CO concentrations observed in Fairbanks have exceeded the NAAQS, the lack of exposure data makes it impossible for the committee to comment on the actual health impacts in Fairbanks from CO episodes. Nevertheless, given the strong health basis of the NAAQS and the observed ambient CO concentrations in Fair-

---

[4]Eight consecutive quarters with no violations allows the borough to request to be redesignated from nonattainment to maintenance status. Such a request may be submitted to EPA during the autumn of 2002 if no violations occur before then.

banks, the committee finds that there are potential health benefits from lowering exposure to CO.

Health Benefits from Meeting CO Standards

Findings

Evidence in the scientific medical literature indicates that attainment of the ambient CO health standards should decrease morbidity and mortality from heart disease. Epidemiological studies at various locations have confirmed that ambient CO concentrations similar to those observed in Fairbanks are associated with increased hospital admissions and increased mortality. For example, one epidemiological study of elderly patients in eight United States counties reported that daily variations in ambient CO concentrations had a statistically significant impact on daily cardiovascular hospital admissions. Another study found a statistically significant relationship between CO and hospital admissions for congestive heart failure in six United States cities. Evidence also suggests that attainment of the 8-h CO standard should generally decrease morbidity from neurological disease, fetal loss, and childhood developmental abnormalities.

Recommendations

To reduce the potential adverse health effects of CO, the borough will need to continue to make progress in meeting the NAAQS for CO. Public-awareness campaigns regarding the public-health issues associated with ambient CO should be enhanced.

Exposure During Episodes

Findings

Human exposure to CO takes place both indoors and outdoors. Air pollution in buildings can come from a combination of the intake of ambient air and indoor sources. In the case of CO, buildings do not provide protection from high outdoor concentrations because the gas is chemically stable, penetrates freely with infiltration air from the outside, and is not removed by building materials or ventilation systems. Indoor sources of CO (for instance, a faulty furnace, an underground parking garage, a kerosene heater, or a smoker), add to the background concentration from the outdoor air. Therefore, concentrations of CO indoors, where people spend the vast majority of their time, can be even higher than those outside. For these reasons, human exposure to high concentrations of CO is likely during episode conditions in Fairbanks, although insufficient exposure data are available to verify this statement.

Recommendations

Additional efforts should be made to monitor personal exposures to CO in buildings and garages near high-CO areas, outside in residential locations, and in motor vehicles. Sampling of human exposure with personal monitoring, blood carboxyhemoglobin measurements, or breath samples is recommended. Monitoring of CO concentrations in microenvironments and of human exposure to CO should be performed in conjunction with improved ambient monitoring to better characterize the CO problem in Fairbanks.

Copollutants

Findings

CO is often an indicator of several less well-characterized pollutants ("copollutants") that are generated and transported with it, especially airborne particulate matter (PM) smaller than 2.5 microns in diameter ($PM_{2.5}$) and toxic organic air contaminants, such as benzene, 1,3-butadiene, and aldehydes. Many of the efforts taken to reduce CO emissions will also reduce emissions of copollutants and their corresponding health effects.

Recommendations

The expected adverse health effects of exposures to copollutants is another reason to consider enhanced CO controls. To assess the relationship between CO and copollutants in Fairbanks during winter, Alaska and the borough should implement an ambient monitoring program for toxic organic compounds and expand the monitoring of $PM_{2.5}$.

Meteorological Conditions of Primary Concern

Findings

When CO concentrations exceed the health standard in Fairbanks, the ambient temperatures are typically between -20° and 20°F (-28.9° and -6.7°C). The combination of high albedo (reflection of sunlight due to snow cover) and the low solar elevation (the sun remains low in the sky) characteristic of northern latitudes in winter creates little heating of the ground and weak vertical mixing between the surface and overlying air. In Fairbanks, that frequently results in ground-based inversions of considerable strength topped by weaker inversions reaching as high as 1-2 km. Human behavior and motor-vehicle technology further narrow the primary temperatures of concern to 0° to 20°F (-17.8° and -6.7°C). In the 0° to 20°F temperature range, preheating a vehicle's engine with electric plug-in devices is not necessary to ensure that it will start, so drivers do not plug in as often as they do when temperatures are below 0°F. The higher emissions that result from starting nonpreheated vehicles is thought to be the reason why five of the last six episodes exceeding the CO standard occurred in this temperature range.

Recommendations

Air quality management in Fairbanks should focus on the 0° to 20°F (-17.8° to -6.7°C) temperature range. Emissions inventories should be refined and verified and control programs evaluated for their effectiveness with emphasis on that temperature range. In addition, air quality modeling should be developed, conducted, and evaluated for the extreme conditions found in Fairbanks in winter.

I/M Programs in the Fairbanks North Star Borough

The vehicle emissions inspection and maintenance (I/M) program is a central element of the borough's plans to comply with the 8-h ambient CO health standard. Emissions models estimate that improved vehicle testing and increased enforcement provide the largest emissions reductions beyond those attributed to increasingly stringent federal motor-vehicle emissions standards. However, the emissions reductions resulting from the I/M program in the borough have not been evaluated with in-use vehicle emissions data. Moreover, the I/M program could be improved. The issues deemed most important by the committee are inspection frequency, exemptions, testing improvements, remote sensing, and continuing program evaluation.

Frequency and Exemptions

Findings

The 1997 change in the borough's I/M program from annual to biennial inspections lacked technical justification and has resulted in an increase in the rate of vehicles failing inspection. Assuming that the quality of vehicle repairs did not change, the increased failure rate implies that the switch to a longer interval has reduced the benefits associated with the I/M program. Returning from biennial to annual inspections will increase the emissions-reduction benefits of the I/M program and help to ensure that the reduction targets of the program are met.

Fairbanks exempts pre-1975 model-year vehicles from its I/M program, although Anchorage tests model years 1968 and newer. Including older vehicles in the Fairbanks program might help to reduce CO emissions from the onroad fleet. Because of the low rates of failure of newer vehicles, studies have shown a minimal loss of program benefits when newer vehicles (vehicles less than 2 or 4 years old) are exempted from the inspection requirement.

Recommendations

The borough should consider resuming annual inspections. The committee is aware that this may require state legislation. The borough should also expand the coverage of its I/M program to include 1968-1974 model-year vehicles. The current new-car testing exemption is reasonable; it may also be cost-effective, starting with the 2000 model year, to expand the exemption to cover the four most recent model years.

Improvements in Emissions Testing

Findings

It is important that the emissions-test procedure be compatible with vehicle technology. The current two-speed idle test for 1982-1995 model-year vehicles is not capable of identifying many of the problems that will cause a vehicle to emit CO at high rates in advanced emissions-control systems, such as a defective oxygen (O_2) sensor, one of the most common significant emissions-producing defects. The committee notes that the borough incorporated advanced onboard diagnostic (OBDII) testing beginning in July 2001 for 1996-model-year and newer vehicles. The OBDII system uses sensors to monitor and modify the performance of the engine and emissions-control components.

Recommendations

The borough should comprehensively assess emissions-testing methods to determine appropriate inspection procedures for various vehicle technologies. This assessment should consider the use of annual two-speed idle tests for pre-1982 vehicles and biennial or annual testing under driving-load conditions for 1982-1996[5] vehicles. The assessment should also consider the issues associated with using OBDII testing in cold climates. Because of the frequency of O_2-sensor failure, the borough should also evaluate the potential emissions-reduction effectiveness of a mandatory O_2-sensor replacement program for older, high-mileage vehicles and implement such a program if it is found to be effective.

Remote Sensing

Findings

Remote sensing is a noninvasive roadside monitoring technology used to estimate the concentration of CO in the exhaust plume of a vehicle as it passes

---

[5]Although OBDII testing can be performed on 1996-model-year vehicles, the OBDII systems on those vehicles have been found unreliable.

a monitored location. Remote-sensing technologies are used in many areas of the United States to track the distribution of vehicle emissions in the onroad fleet and to evaluate the effectiveness of I/M and other emissions-control programs. Although the technology is not typically deployed in the winter because of the effect of extreme cold and other meteorological factors on the performance of remote-sensing equipment, it can be used for part of the year. Vehicles that are high-emitting during the summer are often high-emitting during the winter, so the use of remote sensing could help the borough to evaluate the success of its I/M programs.

Recommendations

Use of remote-sensing capabilities should be considered for the borough's I/M program, as temperatures and atmospheric conditions permit, to help characterize emissions of the vehicle fleet. A continuing remote-sensing program should be considered to evaluate the potential effectiveness of the I/M program, as is done in other regions. The borough could also consider using remote sensing to identify vehicles that must be tested or vehicles that could be given an exemption.

Ongoing Evaluation of I/M

Findings

The I/M program has been predicted to produce substantial reductions in emissions in the Fairbanks area, but the borough has not actually evaluated the effectiveness of the program. Evaluation is especially needed given the borough's claim that the program is much more effective than the federal performance standard for such a program.

Recommendations

The borough should evaluate the I/M program more rigorously to estimate its emissions-reduction benefits and to identify where improvements to the program are needed. The evaluation should allow the direct comparison of the emissions reductions achieved by the program with those estimated in the borough's attainment plan. It should also look for methods to improve the effectiveness of I/M.

Vehicle Plug-ins

Findings

Cold-start emissions occur during the first minutes of vehicle operation—until the engine warms up and the emissions-control catalyst is operating at full efficiency. As the ambient temperature drops, vehicles take longer to warm up, and cold-start emissions increase. Cold-start emissions are especially a problem in Fairbanks during the winter, contributing an estimated 45% of all motor-vehicle emissions. The borough is making substantial efforts to characterize and control these emissions, despite the difficulty in quantifying emissions-reduction credits in its CO-attainment plan.

Electrical heating devices known as plug-ins preheat the engine coolant or lubricant in parked motor vehicles. Plug-ins reduce the amount of time that an engine takes to warm up, reduce fuel consumption, and reduce the length of time needed for the catalyst to become fully operational. Engine preheating can substantially reduce CO emissions during the cold-start phase of engine operations, as well as warming the rider compartment more quickly. The borough is now mandating that parking lots of major employers be equipped with electrical outlets for preheating and is conducting mass-media campaigns to encourage the use of plug-ins during winter.

For plug-ins to be effective in CO control, an electric outlet must be present and operational at the parking space, and the driver must take the time and effort to plug the vehicle in. In the temperature range of 0°F to 20°F (-17.8° to -6.7°C), drivers may make a convenience decision and not plug in if they believe that the temperature will not go below 0°F. Hence, human behavior is a major factor in the effectiveness of this voluntary program. Plug-in use is not being monitored systematically in Fairbanks to demonstrate its emissions-reduction benefits.

Recommendations

The borough should continue to expand the plug-in program by requiring or encouraging the equipping of more parking spaces with electric outlets for plug-ins. Efforts to increase the use of plug-ins at 0° to 20°F (-17.8° to -6.7°C) are especially warranted. Public-education campaigns should continue. Adoption and enforcement of engine-preheating regulations on days expected to have high ambient CO concentrations should be considered. However, further analyses could determine the factors that motivate the voluntary use of plug-ins and the incentives that will expand their use. Additional effort should be directed toward understanding the relationships among engine size, heater power, and the heating time required to substantially reduce cold-start emissions.

Fuel Sulfur Content

Findings

Modern vehicles are equipped with three-way catalytic converters that reduce emissions of organic compounds (including unburned fuel and air toxics), CO, and nitrogen oxides from the engine. High concentrations of sulfur in gasoline decrease the efficiency of the catalytic converter and impair its ability to oxidize CO to carbon dioxide. Switching back and forth between high- and low-sulfur fuels reduces the effectiveness of low-sulfur gasoline.

Two refineries supply the gasoline sold in the Fairbanks region—one with a reported sulfur content of about 200 parts per million (ppm) and the other with a reported sulfur content of less than 1 ppm. The refinery supplying the high-sulfur gasoline is expected to meet the federal regulations for lowering average sulfur content to 180 ppm in 2004 and 30 ppm in 2007. Requiring year-round sale of low-sulfur gasoline in Fairbanks earlier than the federal mandate would help in attaining the CO health standard.

Recommendations

The borough should consider requiring the sale of low-sulfur gasoline as soon as possible. Introduction of lower-sulfur gasoline could be facilitated through accelerated approval of refinery-construction permits and through a state-brokered gasoline-exchange program. Policy and economic analyses, in consultation with the two local refiners, are needed to determine the best approach to ensure that this mandate will not substantially increase the cost of gasoline to Fairbanks residents or compromise the air quality in other parts of the state. A public-awareness campaign to explain the benefits of low-sulfur fuels is needed, and the sulfur content of fuels should be posted at gasoline stations.

Oxygenated Fuels

Findings

Reductions in CO and in some other toxic emissions from oxygenated fuel[6] have been observed at ambient temperatures down to 30°F (-1.1°C). Oxygenated fuels may also provide benefits at temperatures below 30°F, but the effects are uncertain because there has not been much testing in cold climates. Use of oxygenated fuels is highly controversial in the Fairbanks North Star Borough. The decision to use fuels containing the oxygenate methyl *tertiary*-butyl ether (MTBE) in October 1992 was rescinded because of public concerns about odor and possible health effects. There is a reluctance to introduce other oxygenated fuels, such as gasoline containing 10% ethanol, even though such fuels are mandated for Anchorage and estimated by the state to yield the greatest emissions reductions there.

Recommendations

Alaska, EPA, and others should conduct additional research and vehicle testing to assess the effectiveness of ethanol in gasoline for decreasing CO and air-toxics emissions during cold starts and operation at ambient temperatures below 20°F (-6.7°C). If such research indicates that substantial benefits can be achieved, ethanol blending should be considered for Fairbanks.

Traffic Flow and Motorist-Directed Control Strategies

Findings

The borough has implemented several control measures directed at improving traffic flow and modifying motorists' travel behavior, including high-

---

[6]An oxygenated fuel is a gasoline containing oxygenates, typically MTBE or ethanol, intended to reduce production of CO.

way and intersection improvements, transit-system expansion and free transit services, an "alert-day" program, and a public-information campaign. The borough is working on an improved traffic-signal coordination plan. Those measures have contributed to improvements in air quality. Most other transportation control measures (TCMs) recommended by EPA for other areas are clearly not appropriate for Fairbanks given the low levels of congestion, low-density development, and severe winter conditions. It may be possible to achieve marginal reductions in ambient CO, beyond those achieved by plug-ins, with creative and innovative TCMs designed for Fairbanks. However, without sufficient household travel data or a travel-demand forecasting model, the borough does not have the information needed to evaluate the potential effectiveness of alternative TCMs.

Recommendations

The borough should explore parking pricing, telecommuting, and teleservices strategies. The borough should evaluate the effectiveness of its "alert-day" program and consider enhancing it. In addition, a travel-demand study, including a winter travel diary and transit-ridership survey, should be undertaken to provide a basis for evaluating the potential effectiveness of proposed TCMs.

Improving Ambient Monitoring in the Borough

Findings

Ambient air monitoring data are available to characterize long-term CO trends over a limited area in the city of Fairbanks but are inadequate to understand the temporal trends across the entire city and at different heights above the ground. Such data are needed to quantify source contributions to human exposure and to evaluate air quality models.

Recommendations

In the short term, the ambient-CO monitoring network in the borough should be expanded to measure concentrations over a wider area. In the longer term, the vertical distributions of CO concentrations and the wind field should be characterized to support the development and application of modeling approaches better than those now available.

Improving Ambient-CO Modeling in the Borough

Findings

Air quality models are important tools for air quality planning and for forecasting severe CO events. A variety of modeling techniques have been used in other areas to plan strategies for controlling ambient CO, but the extreme conditions in Fairbanks limit the applicability of many of those techniques. In developing Fairbanks's most recent state implementation plan (SIP), Alaska used a relatively simple model, referred to as a statistical rollback model, to estimate the emissions reductions needed to lower ambient CO to achieve the CO health standards. More sophisticated tools could not be used because meteorological data and emissions inventories were insufficient.

Recommendations

In the near term, Alaska should use a simple box-model approach, which simulates the effects of emissions and meteorology in a well-mixed, controlled volume, for air quality planning purposes in Fairbanks. Such an approach could provide greater insights into the effects of the timing of CO emissions and meteorological variables. The relative contributions of mobile and stationary sources to CO episodes could also be assessed with this type of model. Enhanced data-collection efforts are required to support this and more sophisticated modeling efforts.

Improvements in the statistical forecasting approach used by the borough might help in forecasting episodes of high CO concentrations. More work is also needed to develop, apply, and evaluate more sophisticated, physically comprehensive models that would simulate how CO concentrations vary with time and space over the entire borough. Such models could be used for planning, forecasting, and assessing human exposure to high CO concentrations. It is important that model development and testing be specific to the extreme conditions that occur in Fairbanks. Model development must occur in concert with improved monitoring to enable model evaluation.

Public Education

Findings

Successful control of ambient CO requires a community effort. The borough's public-education efforts, aimed at increasing awareness of the CO problem and of how using plug-ins and mass transit may help to alleviate it, have included paid television and radio announcements during heavy viewing and listening times. The committee is concerned, however, that the public-education campaign has not sufficiently emphasized the potential health effects associated with high ambient-CO exposure.

Misconceptions about the relationship between high CO concentrations and public health can be a major barrier to the success of air quality improvement strategies. Policy-makers and the public should understand the health benefits of air quality improvement and accept the need to implement, enforce, participate in, and support these strategies.

Recommendations

Public-education programs should be continued and expanded to increase public awareness of the potential health effects of high ambient CO concentrations and to increase public participation in efforts to improve air quality. Surveys of public opinion should be used in designing the programs and assessing their effectiveness.

OVERALL EVALUATION

Fairbanks has succeeded in reducing the number of days that violate the 8-h CO standard from over 130 per year in the 1970s to none during the last 2 years. Despite this improvement, it is likely that, under severe inversion conditions, Fairbanks will again exceed the 8-h CO standard, particularly if the area experiences significant growth due to pipeline or military construction. To further improve air quality in Fairbanks, the borough should continue to use, and possibly expand, effective current control strategies. The committee concludes that pursuit of cold-start emissions controls through the plug-in program should continue to be a priority. For its final report, the committee will consider the extent to which the federal government could aid Fairbanks and other areas in improving air quality by mandating stricter vehicle-certification standards for starts during cold conditions. The committee sees such standards as possibly a very effective means for the borough to reduce emissions in the future but also recognizes that national vehicle-emissions standards cannot be set for a single area.

In addition, the borough should undertake a cost-effectiveness analysis to help determine which of the other emissions-control strategies discussed in this report should be pursued. Efforts to control CO in the borough should be accompanied by public-education campaigns to improve public awareness of and participation in efforts to improve air quality.

The committee also sees potential benefits from additional studies to improve the spatial characterization of CO during episodes of high ambient concentrations. In the short-term, continuing to conduct a basic assessment of the spatial variability of CO during such episodes should be a priority. The long-term priority should be to develop a three-dimensional characterization of CO episodes; that would require long-term monitoring commitments and the development of modeling capabilities. Such work would probably require local and state commitments in combination with federal research support.

Further study of the Fairbanks CO problem could provide useful insights to scientists and regulators in the wider air quality community. Fairbanks constitutes a natural laboratory for understanding influences of meteorology and topography on air quality in regard to CO and for understanding the effectiveness of emissions-control technologies at low temperatures.

1

Sources and Effects of Carbon Monoxide Emissions

THE PROBLEM

Carbon monoxide (CO)—a colorless, odorless, tasteless, and toxic air pollutant—is produced in the incomplete combustion of carbon-containing fuels, such as gasoline, natural gas, oil, coal, and wood. The largest anthropogenic source of CO in the United States is vehicle emissions. Breathing the high concentrations of CO typical of a polluted environment leads to reduced oxygen (O_2) transport by hemoglobin and has health effects that include headaches, increased risk of chest pain for persons with heart disease, and impaired reaction timing. In the 1960s, vehicle emissions led to increased and unhealthful ambient CO concentrations in many U.S. cities. With the introduction of emissions controls, particularly automotive catalysts, estimated CO emissions from all sources decreased by 21% from 1980 to 1999 (EPA 2001a). Average ambient concentrations decreased by about 57% over the same period (EPA 2001a).

The locations that continue to have high concentrations of CO tend to have topographical or meteorological characteristics that exacerbate pollution; for example, strong temperature inversions or the existence of nearby hills that inhibit wind flow may limit pollutant dispersion. Because of the limited dispersion, many of those areas also have unhealthful concentrations of summer ozone (O_3) and year-round particulate matter (PM).[1] Low temperatures also contribute to high CO concentrations. Engines and vehicle emissions-control equipment operate less efficiently when cold: Air-to-fuel ratios are lower, combustion is less complete, and catalysts take longer to become fully operational. The result is that products of incomplete combustion, including CO, are formed in higher concentrations. Sometimes, topography, meteorology, and emissions combine to cause high concentrations of CO. Compliance with the health-based National Ambient Air Quality Standards (NAAQS) for CO has proved difficult under those circumstances. The question arises as to whether unique methods are necessary to manage CO in such problem areas, or whether the current policies will ultimately achieve good air quality.

CHARGE TO THE COMMITTEE

In response to the challenges posed for some areas by having to come into compliance with the NAAQS for CO, a committee was established by the National Research Council (NRC) to investigate the problem of CO in areas with meteorological and topographical problems. The committee's statement of task is as follows:

An NRC committee will assess various potential approaches to predicting, assessing, and managing episodes of high concentrations of CO in meteorological or topographical problem areas. The committee will consider interrelationships among emissions sources, patterns of peak ambient CO concentrations, and various CO emissions-control measures in such areas. In addition, the committee will consider ways to better understand relationships between episodes of high ambient CO and personal exposure, the public-health impact of such episodes, and alternative ways to measure progress in controlling ambient CO. An interim report dealing with Fairbanks, Alaska, as a case study is to be completed. A final report, including other CO problem areas, will be completed by the end of the study.

---

[1] Of the seven areas currently in serious nonattainment for CO, five (Anchorage, Denver-Boulder, Los Angeles, Phoenix, and Spokane) are also in nonattainment for PM_{10}, and three (Denver-Boulder, Los Angeles, and Phoenix) are in nonattainment for ozone (O_3). O_3 formation is promoted by sunlight and high temperatures and is a summer problem.

The committee will address the following specific issues:

- Types of emissions sources and operating conditions that contribute most to episodes of high ambient CO.
- Scientific bases of current and potential additional approaches for developing and implementing plans to manage CO air quality, including the possibility of new catalyst technology, alternative fuels, and cold-start technology as well as traffic and other management programs for motor-vehicle sources. Control of stationary source contributions to CO air quality will also be considered.
- Assessing the effectiveness of CO emissions-control programs, including comparisons among areas with and without unusual topographical or meteorological conditions.
- Relationships between monitored episodes of high ambient CO concentrations and personal human exposure.
- The public-health impact of such episodes.
- Statistically robust alternative methods to assist in tracking progress in reducing CO that bear a relation to the CO concentrations considered harmful to human health.

This study is expected to provide scientific and technical information potentially helpful to the development of state implementation plans; however, the committee will not provide prescriptive advice on the development of specific state implementation plans for achieving CO attainment. In addition, it will not suggest changes in regulatory compliance requirements for areas in nonattainment of the NAAQS, and it will not recommend changes in the NAAQS for CO.

Fairbanks, Alaska, was chosen as a case study for this interim report because its meteorological and topographical characteristics make it susceptible to severe winter inversions that trap CO and other pollutants near ground level. The combination of low temperatures and low solar radiation during the winter decreases the rate at which CO is ventilated from the area, while Fairbanks's location in a river valley with hills on three sides further inhibits pollutant dispersion. In addition, Fairbanks experiences relatively high winter emissions per vehicle because of its low temperatures.

In this interim report, the committee addresses those parts of the charge pertinent to Fairbanks. Specifically, the committee reviewed Fairbanks in terms of the meteorological and topographical conditions that foster pollution episodes, CO emissions from mobile sources in severe winter conditions, air quality management options, and statistical methods for tracking progress. Because no exposure data were available in Fairbanks and the small population size makes epidemiological studies difficult, the committee was unable to directly evaluate the public-health impact of high ambient CO concentrations there. Nevertheless, after reviewing the evidence on the health effects of CO presented in the literature and summarized by EPA in their most recent criteria document (EPA 2000a), the committee agrees that meeting the NAAQS for CO will protect human health with an adequate margin of safety. It is important to note that the committee was specifically instructed not to suggest changes to the levels or the form of the NAAQS for CO.

The final report, which will be issued in early 2003, will look at the characteristics of the CO problem in other meteorological and topographical problem areas and at the applicability of the various CO-control measures in these areas. One subject of particular interest will be the extent to which areas other than Fairbanks would benefit from stricter vehicle-certification standards that control emissions produced by vehicle starts at low temperatures.

HEALTH EFFECTS OF CO

Clinical and Epidemiological Studies of CO Effects

CO affects human health by impairing the ability of the blood to bring O_2 to body tissues. When CO is inhaled, it rapidly crosses the alveolar epithelium to reach the blood, where it binds to hemoglobin to form carboxyhemoglobin (COHb), a useful marker for predicting the health effects of CO. Because CO has an affinity for hemoglobin more than 200 times greater than does O_2, the presence of CO in the lung will displace O_2 from the hemoglobin. In other words, when CO is present in the lungs, the hemoglobin will be unable to reach 100% O_2 saturation. In addition, the presence of COHb increases hemoglobin's affinity for O_2, thereby inhibiting release of O_2 from the hemoglobin to body tissues. The effect of COHb is illustrated by a leftward shift of the O_2-hemoglobin dissociation curve (Figure 1-1). As shown in Figure 1-1, once COHb forms, the hemoglobin is unable to reach 0% O_2 saturation. This second effect continues until the COHb dissociates, typically several hours after CO exposure. CO not only decreases the O_2-carrying capacity of the blood, but also decreases the ability of the tissues to extract O_2 from the blood during circulation. CO has also been shown to bind to myoglobin and may affect O_2 transport to muscle (EPA 2000a).

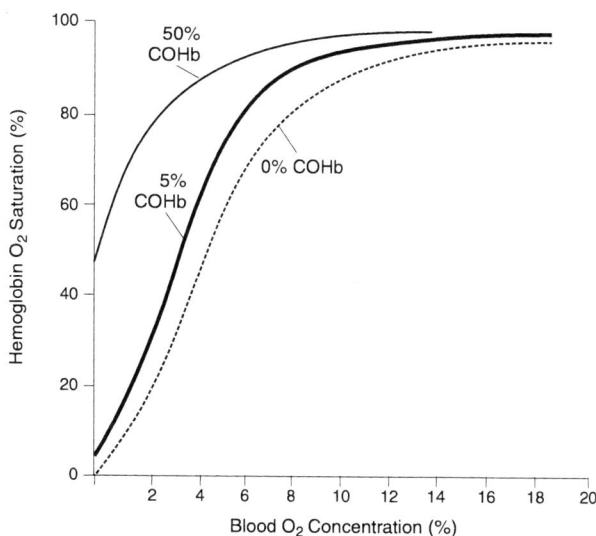

FIGURE 1-1 Diagram of hemoglobin response to the presence of COHb. The concentration of O_2 in the environment surrounding the hemoglobin is shown on the x-axis. The O_2 saturation, or how much of the hemoglobin's capacity for storing O_2 is used, is shown on the y-axis. At higher O_2 concentrations, as are found in the lungs, the hemoglobin can be more O_2 saturated. Likewise, at lower O_2 concentrations, as are found in other parts of the body, O_2 will dissociate from the hemoglobin to achieve O_2 saturations as indicated by the curve. The presence of COHb shifts this curve to the left. For a given O_2 concentration, the hemoglobin will require a higher O_2 saturation and allow less O_2 to be released to body's tissues. Source: Adapted from Shephard 1983.

COHb levels in healthy individuals not recently exposed to high concentrations of ambient CO are 0.3 to 0.7%. Exposure to high concentrations of ambient CO can result in concentrations of COHb of 2% or higher if the exposure lasts long enough (hours). For those who smoke, cigarette-smoking is typically the most significant source of personal CO exposure. COHb concentrations, which are generally less than 1% in nonsmokers, average about 5% in smokers and are up to 10% or even higher in some very heavy smokers (Beckett 1994).

The CO health standards set by EPA are intended to keep COHb concentrations for nonsmokers below 2% in order to protect the most susceptible members of the population. EPA last published the *Air Quality Criteria for*

Carbon Monoxide, which summarizes health research findings and their implications for setting the NAAQS, in June 2000 (EPA 2000a). That document provides a comprehensive review of the literature pertaining to the health effects of CO for typical environmental exposures that would be associated with COHb levels less than 10%. The major findings presented by EPA (EPA 2000a) are summarized below; the committee accepts these findings as sufficient evidence of the health effects caused by exposure to CO at concentrations of 9 parts per million (ppm) and above for an extended period of time.

The acute affects of CO poisoning are well understood (Raub et al. 2000). Generally, in otherwise healthy people, headache develops when COHb concentrations reach 10%; tinnitus (ringing in the ear) and lightheadedness at 20%; nausea, vomiting, and weakness at 20-30%; clouding of consciousness and coma at around 35%; and death at around 50% (Coburn 1970). However, the outcomes of long-term, low-concentration CO exposures are less well understood. Because of the critical nature of blood flow and O_2 delivery to the heart and brain, these organ systems, as well as the lungs (the first organ to come into contact with the pollutant), have received the most attention.

In patients with known coronary artery disease, COHb concentrations as low as 3% exacerbate the development of exercise-induced chest pain (Allred 1989a). Concentrations as low as 6% are associated with an increase in the number and frequency of premature ventricular contractions during exercise in patients with severe heart disease (Allred 1989b; Sheps et al. 1990). Large environmental-exposure cohort studies have confirmed that daily increases in ambient CO concentrations are associated with statistically significant increases in the numbers of hospital admissions for heart disease (Poloniecki et al. 1997; Schwartz 1999) and congestive heart failure (Morris et al. 1995) and with increases in deaths from cardiopulmonary illnesses (Prescott et al. 1998).

Neuropsychiatric (neurological and psychiatric) disorders and cognitive impairments due to long-term, low-concentration CO exposures have been hypothesized in part on the basis of extrapolation from the known acute effects of high-dose CO poisoning and the concomitant subacute and delayed neuropsychological sequelae. In clinical experiments on healthy volunteers, controlled CO exposure was associated with subtle alterations in visual perception when COHb concentrations were above 5% (McFarland 1970; Horvath et al. 1971). However the significance of this finding remains unknown. Similar studies have shown measurable but small effects on auditory perception, driving performance, and vigilance (Beard and Wertheim 1967; McFarland 1973; Benignus et al. 1977).

The role of CO in pulmonary disease is unclear. In the Seattle area, a single-pollutant model showed a 6% increase in the rate of hospital admissions for asthma with each 0.9-ppm increase in CO, but that was concomitant with increases in other air pollutants (Sheppard et al. 1999). In Minneapolis and Toronto, CO concentrations showed only weak and inconsistent associations with total admissions for respiratory diseases (Burnett et al. 1997; Moolgavkar et al. 1997).

A fetus is more susceptible to CO than an adult; the O_2-hemoglobin dissociation curve is to the left of that in the adult. It is shifted even further to the left by CO exposure. Also, because the half-life of fetal COHb is longer than that of adults, it may take up to five times longer to reduce the concentrations to normal. Other studies have shown that exposure to high concentrations of CO during the last trimester of pregnancy may increase the risk of low birth weights and that exposures to CO and PM during pregnancy may trigger preterm births (Ritz and Yu 1999; Ritz et al. 2000). A recent study linked CO and O_3 exposure during pregnancy to birth defects such as cleft lip and defective heart valves (Ritz et al. 2002). However, the lead author of the study cautioned that the real culprit may be other pollutants that are dispersed with CO in tailpipe emissions (Beate Ritz, personal communication, January 31, 2002).

Public-health laws are designed to protect the most susceptible people in the population. People with coronary artery disease or other cardiopulmonary diseases, fetuses, infants, and athletes who exercise heavily in high-CO atmospheres are particularly susceptible to experiencing adverse health effects from CO. The evidence summarized above, and described more fully by EPA (2000a), indicates that attainment of the ambient-CO standards can decrease morbidity and mortality from atherosclerotic heart disease. Although less conclusive, there is evidence that attainment of the CO standards will also decrease morbidity from pulmonary disease, neurological disease, fetal loss, and childhood developmental abnormalities. These health benefits translate into economic savings associated with avoided health care and avoided work-time losses as well as intangible savings in life quality.

CO Exposure

Motor-vehicle emissions are the primary source of CO in outdoor air in populated areas and are associated with the highest outdoor CO exposure in nonsmokers. Outdoor concentrations of CO tend to be higher in urban areas and to increase with the density of vehicles and miles driven. Measurements of ambient CO typically exhibit a bimodal diurnal pattern, with the highest concentrations generally occurring on weekdays during the commuting hours of 7:00-9:00 a.m. and 4:00-6:00 p.m. (EPA 2000a).[2] CO also accumulates in the rider compartments of motor vehicles. Studies have shown that when the concentration near roadways averages 3-4 ppm, the average concentration in the cab is typically 5 ppm (Akland et al. 1985; Flachsbart et al. 1987).

Most people spend a majority of their time indoors; this is particularly true in Fairbanks and other cold climates during the winter, when ambient CO concentrations tend to be highest. That leads to the question of the relationship between indoor and outdoor concentrations. Air pollution in buildings can come from indoor sources and from air exchange with outdoor ambient pollution. Air exchange may be active, as in the case of a mechanical ventilation system, or passive, as in the case of infiltration associated with temperature or pressure differences between the outside and the interior of a building. CO penetrates freely with infiltration air from the outside and is not removed by building materials or ventilation systems. Furthermore, there are no effective indoor chemical or physical processes for lowering CO on the time scales of interest for exposure and toxic effects.

The relationship between indoor and outdoor CO concentrations can be evaluated with a simple differential mass-balance model (Shair and Heitner 1974) that has the following steady-state solution when we combine active ventilation and passive infiltration into a single air-exchange term:

$$C_i = \frac{paC_o}{a+k} + \frac{S}{(a+k)V}, \qquad (1)$$

where

C_i = indoor concentration, µg/m³;
C_o = outdoor concentration, µg/m³;
p = penetration coefficient, 0-1;
a = air exchange rate, h⁻¹;
k = decay rate, h⁻¹;
S = mass flux of the indoor source, µg/h; and
V = building volume, m³.

---

[2]Although most areas show a bimodal diurnal pattern with respect to ambient CO concentrations, Fairbanks, Alaska, the focus of this interim report, typically shows a continuous increase throughout the day, with 1-h average CO concentrations peaking at 5:00-6:00 p.m. (EPA 2000a).

For CO, the relationship is simpler because the penetration coefficient (p) is unity and the decay rate (k) is effectively zero. Therefore, the solution is

$$C_i = C_o + \frac{S}{aV}. \qquad (2)$$

It is clear that in the absence of indoor sources (S), the steady-state indoor concentration of CO will equal the outdoor concentration. When a source of CO is present indoors (for example, from a faulty furnace, an underground parking garage, a kerosene heater, or a smoker), the indoor source adds to the background concentration from the outdoor air (EPA 2000a). Therefore, buildings do not provide protection from high outdoor concentrations of CO. The idea that buildings provide protection from high outdoor CO concentration is a common misconception.[3]

Related Pollutants

The incomplete combustion of fossil fuels, which is responsible for CO emissions, also causes emissions of fine particles ($PM_{2.5}$) and toxic organic air contaminants. Epidemiological studies have linked exposure to $PM_{2.5}$ with various adverse health effects, including premature mortality, exacerbation of asthma and other respiratory tract diseases, and decreased lung function (NRC 1998; EPA 2001b). Because of these adverse health effects, EPA issued NAAQS regulating ambient concentrations of $PM_{2.5}$ (EPA 1997). In addition to the six criteria air pollutants previously regulated, the 1990 amendments of the Clean Air Act (CAAA90) designated 189 toxic air contaminants. Incomplete combustion in mobile sources is estimated to contribute a substantial fraction to the emissions of several toxic air pollutants, including benzene, 1,3-butadiene, and aldehydes (EPA 2001c). Each of these toxic air pollutants poses some carcinogenic risk. In addition, chronic exposure to benzene is associated with blood disorders; chronic exposure to 1,3-butadiene is associated with cardiovascular disease; and chronic exposure to aldehydes is associated with respiratory problems and eye, nose, and throat irritation (EPA 1994). Many emissions-control strategies for CO will also cause reductions in these copollutants and their associated adverse health effects.

---

[3]In this regard, CO is different from O_3, which is highly reactive and therefore rapidly destroyed when infiltrating inside from outdoors.

CO may be a good indicator gas for other pollutants that are emitted at the same time but are not widely measured. The concentrations and spatial distributions of the copollutant species are generally not as well known as for CO. In particular, little data are available about exposure to air toxics present in the ambient environment. CO could be especially useful as an indicator of mobile-source emissions of $PM_{2.5}$ and air toxics, which some studies have shown to be strongly correlated with CO. In one study of emissions from in-use vehicles sampled in Denver, San Antonio, and the Los Angeles area, strong correlations were found between CO and particle emissions ($R^2 = 0.65$) and between particle and total hydrocarbon (HC) emissions ($R^2 = 0.78$) (Cadle et al. 1999). The same study demonstrates that emissions of the pollutant species increase with vehicle age and during cold starts. For individual vehicles, however, the correlation among the pollutants is weaker, reflecting the complex mechanisms of formation of the related combustion products.

In another study, three vehicles from Fairbanks were evaluated with the Federal Test Procedure (FTP) for the Urban Dynamometer Driving Schedule (Mulawa et al. 1997). Particle emissions were higher at lower temperatures; the average PM emissions from the three tested automobiles (using regular gasoline) increased from 14.2 to 44.2 mg/mile as the temperature decreased from 20°F to -20°F. Most of the particles were released during Phase I of the FTP, representing the first 505 seconds during cold start. The particle emissions also correlated with HC and CO emissions and appeared to be related to rich-operating conditions. Mechanisms of PM formation may include oil consumption.

NATIONAL AIR QUALITY STATUS FOR AMBIENT CO

NAAQS

In recognition of the adverse health effects of CO, the U.S. Environmental Protection Agency (EPA), as directed by the Clean Air Act (CAA), established health-based air quality standards for CO in 1971. Recognizing that exposure can have both acute and chronic effects, the primary NAAQS for CO has two criteria with different averaging periods: 35 ppm averaged over 1 hour (h), and 9 ppm averaged over 8 h. Each standard is not to be exceeded more than once per year; the second exceedance and each subsequent exceedance

> **BOX 1-1** Recommendations: Health and Exposure
>
> *Health Benefits from Meeting CO Standards*
>
> To reduce the potential adverse health effects of CO, the borough will need to continue to make progress in meeting NAAQS for CO. Public-awareness campaigns regarding the public-health issues associated with ambient CO should be enhanced.
>
> *Exposure During Episodes*
>
> Additional efforts should be made to monitor personal exposures to CO in buildings and garages near high-CO areas, outside in residential locations, and in motor vehicles. Sampling of human exposure with personal monitoring, blood carboxyhemoglobin (COHb) measurements, or breath samples is recommended. Monitoring of CO concentrations in microenvironments and of human exposure to CO should be performed in conjunction with improved ambient monitoring to better characterize the CO problem in Fairbanks.
>
> *Copollutants*
>
> The expected adverse health effects of exposures to copollutants is another reason to consider enhanced CO controls. To assess the relationship between CO and copollutants in Fairbanks during winter, Alaska and the borough should implement an ambient monitoring program for toxic organic compounds and expand the monitoring of $PM_{2.5}$.

within a year are considered violations of the standard.[4] The standards have been periodically reviewed on the basis of new scientific findings as mandated by the CAA. The most recent review was published in 2000, when the standards were reaffirmed (EPA 2000a).

---

[4]For 8-h averages, the violation occurs if there are at least two nonoverlapping 8-h time periods, each with average CO concentration above 9 ppm.

The 8-h standard of 9 ppm is the more difficult to attain. EPA originally designated an area as "nonattainment" if the second highest 8-h CO concentration measured during a calendar year (termed the "design value") was greater than 9 ppm. After the CAAA90, the EPA administrator designated each area that had previously been in nonattainment as "serious" if the design value was 16.5 ppm or greater, "moderate" if the design value was 9.1-16.4 ppm, "not classified" if recent data were insufficient to determine whether the standard was met, or in attainment. Moderate areas that did not reach attainment by July 1996 could be reclassified as serious by EPA. Each nonattainment area is required to submit to EPA a state implementation plan (SIP) that includes a characterization of pollutant concentrations and emissions, a description of the emissions reductions the area plans to make, and an "attainment demonstration" showing how the emissions reductions will enable the area to attain and maintain compliance with the NAAQS. To be eligible to request reclassification from nonattainment to attainment status, an area must have air quality monitoring data to show that it did not violate the standard for the previous 2 years (y).

Areas in Nonattainment for CO

Following the CAAA90, 42 areas were designated as nonattainment for the 8-h standard for CO. As of this time, seven areas remain in serious nonattainment status for the 8-h standard, although only three areas failed to meet the CO NAAQS in 1999-2000: Fairbanks, Alaska; Lynwood, California, in the South Coast Air Basin; and Calexico, California, near the Mexican border (EPA 2000b). All areas are in attainment with respect to the 1-h standard. Monitoring in Fairbanks in 2000 and 2001 did not show an exceedance, so the borough is now eligible to apply for attainment status. Lynwood apparently suffers from an older vehicle fleet with higher emissions and inhibited dispersion caused by strong inversions combined with low windspeeds at night (Bowen et al. 1996). Calexico suffers from cross-border pollutant transport and is not considered to be primarily a meteorological or topographical problem area. Continued reductions in emissions are expected in Fairbanks and Lynwood over the next decade as newer, cleaner vehicles replace older ones. However, unusual meteorological conditions or unexpected increases in emissions (in the case of Fairbanks, construction of a trans-Alaska natural-gas pipeline or substantial missile defense facilities) could prevent those areas from maintaining compliance.

The seven areas classified as being in serious nonattainment as of 2001 are listed in Table 1-1. In addition, for each area, the year from 1995 to 2000 that had the most exceedances is shown with the highest and second-highest 8-h average CO concentrations measured during that year. The highest 8-h average CO concentrations occurred early in the 6-y period in each area; this is consistent with the general decline in CO concentrations across the country during that period. The highest number of exceedances and the highest 8-h averages are associated with the Los Angeles South Coast Basin, and the Lynwood area in particular. Fairbanks ranks second and is unusual for such a small metropolitan area in having such a serious air quality problem. CO concentrations measured in Calexico, California, exceed the 8-h standard, but that area has not yet been classified by EPA.

METEOROLOGY AND TOPOGRAPHY IN CO NONATTAINMENT AREAS

When ventilation in a region with high CO emissions from a nearby source is restricted, CO can accumulate in the air near the ground. The air can be trapped vertically, by temperature inversions, and horizontally, by topography or stagnant meteorological conditions. Stagnation is characterized by very low windspeeds. Figure 1-2a,b illustrates air-stagnation cases in the continental United States (Wang and Angell 1999). The Southwest and parts of the Gulf Coast are clearly susceptible to air stagnation.

Vertical ventilation depends on how air temperature varies with altitude. The average temperature in the troposphere decreases with altitude at about 6.5°C/km, as shown in Figure 1-3 (Huschke 1959). Under some meteorological conditions, temperature in the lower atmosphere may increase with altitude for short distances. Such temperature inversions inhibit vertical mixing be cause less-dense warm air rests above colder, denser air. The temperature inversion therefore defines the vertical limit of mixing. For example, as shown in Figure 1-3b, pollutants emitted at the surface could be mixed up to about 600 m. The inversion strength depends on the rate at which temperature increases with altitude; stronger inversions have a more rapid increase of temperature with altitude.

As illustrated in Figure 1-3a,b, inversions can be surface-based and due generally to surface cooling or can be at a high altitude and due generally to horizontal advection of warm air aloft or subsidence (downward motion of air). An existing inversion can be strengthened by atmospheric subsidence in

TABLE 1-1 U.S. Areas Classified As Serious Nonattainment for CO As of 2000

Area Name	Population (1990)	Areas (mi²)	Number of Monitors	Worst Year in 1995-2000	During the Worst Year in 1995-2000		
					Highest 8-h Average CO (ppm)	2nd Highest 8-h Average CO (ppm)	Number of exceedances[a]
Fairbanks, AK City	30,000	32	3	1995	15.2	11.8	9
Anchorage, AK Borough	80,000	7,366					
Anchorage, AK	222,000	43	3	1996	11.0	9.6	3
Las Vegas, NV	258,000	1,320	7	1996	10.3	10.1	3
Spokane, WA	279,000	161	7	1995	13.1	11.2	4
Denver-Boulder, CO	1,800,000	1,500	7	1995	11.0	9.5	2
Phoenix, AZ	2,006,000	2,580	9	1995	10.2	9.9	3
LA-South Coast, CA	13,000,000	6,000	14	1996	17.5	14.5	22

[a]Number of times 8-h average CO concentrations in the area were greater than of equal to the standard of 9 ppm.
Source: Larry Elmore, EPA, Office of Air Quality Planning and Standards.

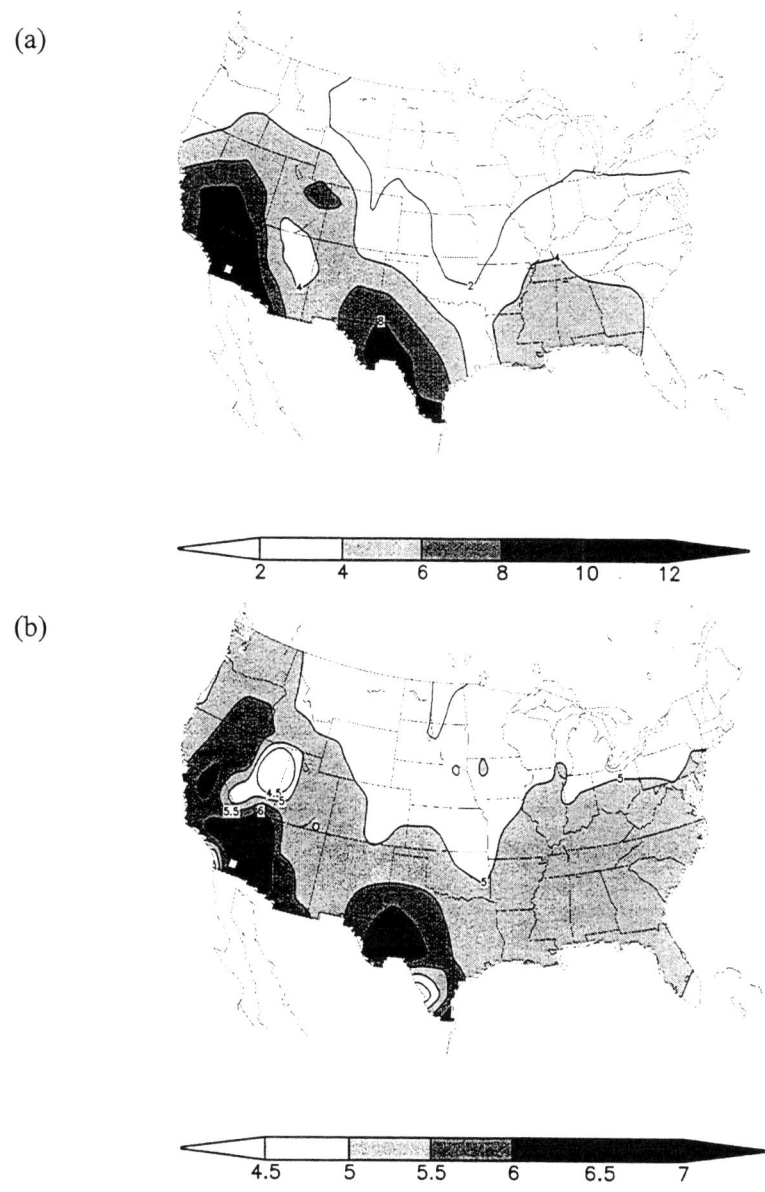

FIGURE 1-2a,b (a) Air stagnation annual mean cases for 1948-1998 for the contiguous 48 states. (b) Mean duration of stagnation cases, in days. Air is considered stagnant if for at least 4 d sea level geostrophic windspeeds are less than 8 m/s, there is no precipitation, and the winds at 500 hPa are less than 13 m/s.

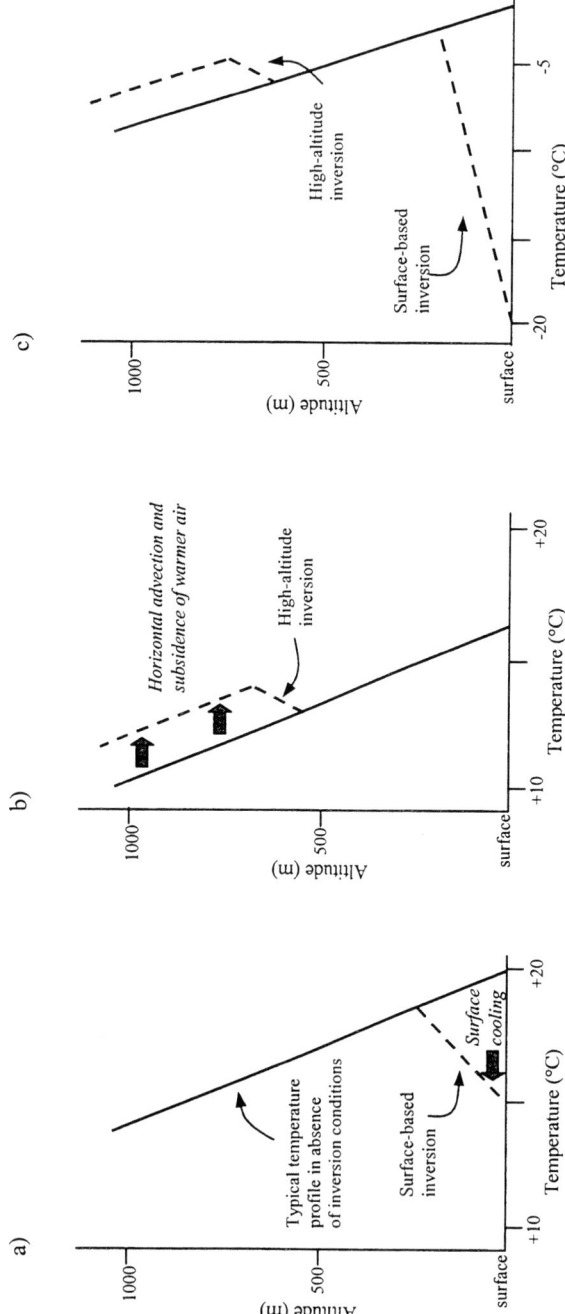

FIGURE 1-3 Schematics of (a) a surface-based inversion, (b) a high-altitude inversion, and (c) a typical Alaskan temperature profile with both surface-based and high-altitude inversions. Solid lines indicate the temperature profile in the absence of inversions. Dashed lines indicate temperature profiles affected by inversions.

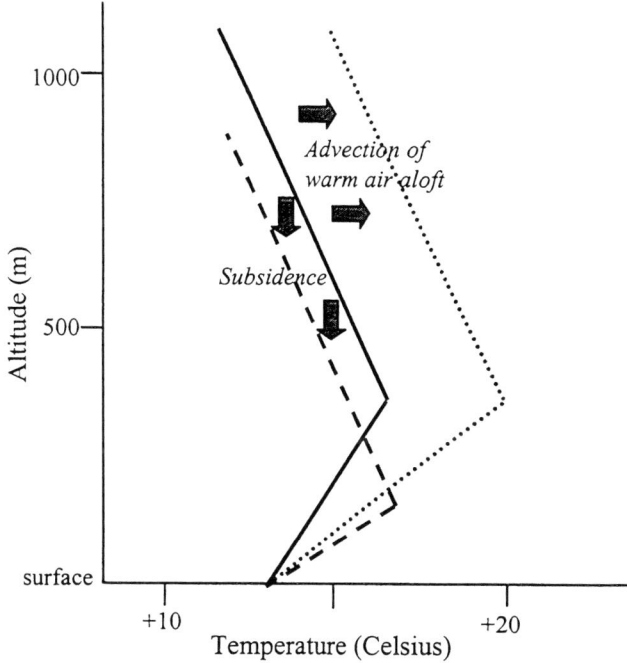

FIGURE 1-4 Schematic of how an existing surface-based inversion (solid line) can be strengthened by subsidence (dashed line) or by advection of warm air aloft (dotted line).

the region or by the transport into the region of air aloft that is warmer than the air at the surface (Figure 1-4). Indeed, regions under high-pressure systems, where subsidence and the advection of warm air aloft can both be present (Figure 1-5), typically experience inversion conditions. Thus, urban locations that have frequent high-pressure systems often suffer from serious air pollution because inversions limit the vertical mixing of pollutants (Pielke et al. 1987; Stocker et al. 1990).

In central Alaska, the combination of clear skies, light surface winds, snow cover, and the absence of solar heating during the winter often produces exceptionally strong surface inversions (Figure 1-3c). Radiation of heat to space produces low surface temperatures that can trap air and pollutants during high-pressure subsidence or warm-air advection from the south. It is likely that warm-air advection aloft from the south is necessary to produce CO exceedances in Fairbanks (S.A. Bowling, Fairbanks Meteorology, in prepara-

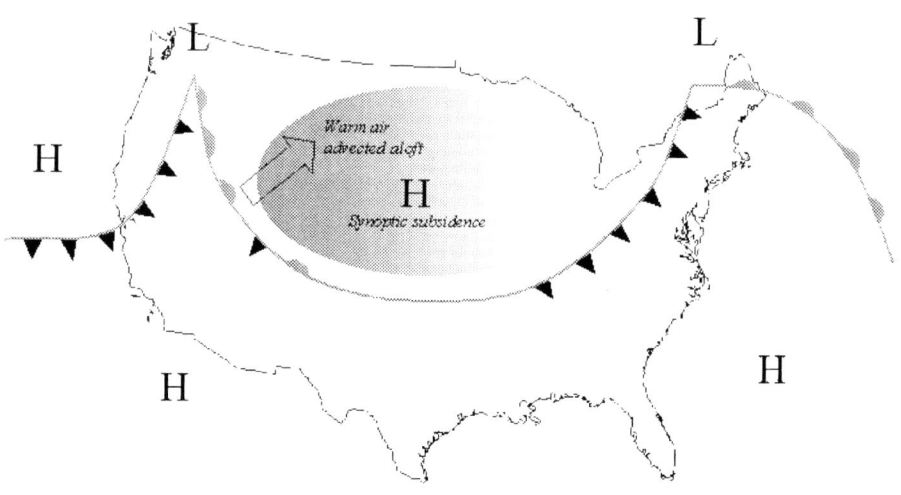

FIGURE 1-5 Schematic of how major synoptic weather features can strengthen inversions. Areas of high pressure (indicated by "H") experience synoptic subsidence. At the same time, warm fronts (indicated by semicircles) advect warm air above the surface. The shaded area indicates a region where existing inversion conditions can be strengthened by both subsidence and advection of warm air aloft.

tion, 2002). In contrast, southwestern California experiences inversions at higher altitudes because of strong subsidence associated with descending air in the subtropical eastern Pacific Ocean high-pressure system. Inversions in southwestern California are high because, in contrast with the winter situation in Fairbanks, sunshine on the surface permits vertical mixing in the atmospheric layer near the surface. Topographical features, such as the presence of an urban area in a valley, can inhibit horizontal dispersion of air. Unless the large-scale wind is strong enough to enable horizontal dispersion, the local air mass will remain confined to the valley (see Figure 1-6). Los Angeles is subject to that problem because it has mountains on three sides. Similarly, Fairbanks has high terrain to its west, north, and east. Because cities tend to develop at lower elevations, often in river valleys, terrain trapping is a concern for many urban areas. Valleys well above sea level have an added problem in that it is easy to lose heat to space at high elevations. Pielke et al. (1991) proposed a method to estimate worst-case air quality in complex terrain.

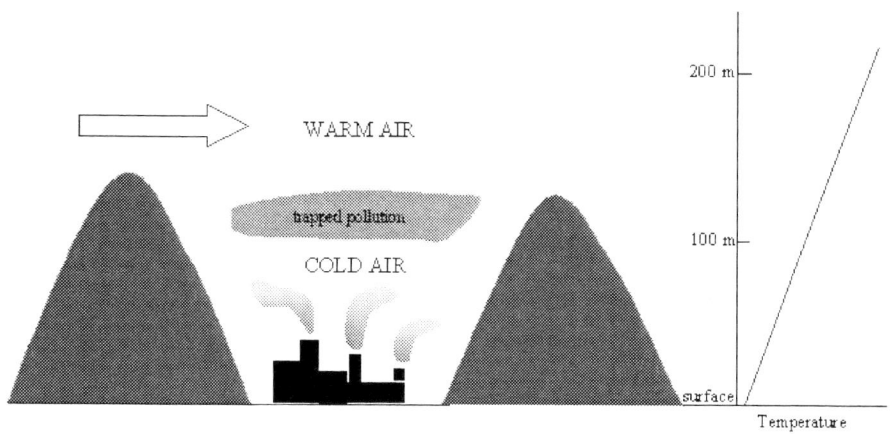

FIGURE 1-6 Schematic of a trapping valley. The temperature profile in the valley is shown on the right.

NATIONAL INVENTORY OF CO EMISSIONS

An emissions inventory is an accounting of all emissions of a pollutant in a defined area in a specified period. It is common to calculate the annual emissions of CO in nonattainment areas; the estimates are used to compare emissions in various years and locations, assess the effectiveness of air pollution policies, and predict the impact of future projects on emissions. Determining an emissions inventory requires that all major sources of a pollutant be identified, that the releases from the sources be quantified, and that the releases be summed to determine the total mass emitted in the defined area. Major sources are usually quantified with the aid of emissions rates, which are multiplied by an activity rate to yield the mass of emissions released.

For example, annual highway vehicle emissions are commonly determined by multiplying the emissions rate, in grams per mile, by the number of miles traveled in a year. Likewise, emissions from oil heaters are usually determined by multiplying the emissions rate, in grams per gallon of heating oil, by the number of gallons of heating oil used.

Table 1-2 is an inventory of CO emissions in the United States during 1999 (EPA 2001a). An estimated 77% of the anthropogenic CO emissions are from mobile sources, including onroad vehicles (51% of the total) and nonroad engines and vehicles (26% of the total). The remaining CO emissions are from area and point sources, including fuel combustion and industrial processes. In urban areas, mobile sources may contribute relatively more or less than the national average to the mix of emissions. For example, mobile sources contributed 89% of CO emissions in Anchorage, Alaska, but contributed only 76% of CO emissions in Denver, Colorado. On the basis of its Mobile Source Emissions Factor (MOBILE) model, EPA suggested that vehicles may contribute 90% or more of CO emissions in cities with serious air pollution (EPA 1993). As described in NRC's report (NRC 2000), however, studies evaluating the MOBILE model have shown substantial inaccuracies in its estimates of fleet emissions and the effectiveness of control strategies.

CO EMISSIONS FROM VEHICLES

The primary source of CO from vehicles is the incomplete combustion of gasoline in engine cylinders. The fuel-oxidation process (combustion) is the conversion of the fuel to lower-molecular-weight intermediate HCs (including olefins and aromatics) and their conversion to aldehydes and ketones, then to CO, and finally to carbon dioxide (CO_2). The initial reactions are faster than the final conversion of CO to CO_2. Incomplete conversion of fuel carbon to CO_2 results in part from insufficient O_2 in the combustion mixture—known as fuel-rich[5] conditions—and insufficient time to oxidize fuel carbon fully to CO_2. CO emissions by diesel vehicles are minimal, primarily because of the excess air used in the diesel combustion cycle. Hence, the following discussion is limited to gasoline vehicles.

Vehicle Technologies

Before the 1980s, carburetors were used to meter fuel in proportion to the intake air of gasoline engines. The design of the carburetor typically ensured

---

[5]The ratio of air to fuel mass that provides just enough O_2 to convert all the carbon and hydrogen in gasoline to CO_2 and water (called stoichiometric) is about 14.7:1. Ratios less than 14.7:1 have more fuel than this optimal mixture and are termed fuel-rich.

TABLE 1-2 National CO Emissions Inventory Estimates for 1999

Source Category	Thousands of Short Tons
Point- or Area-source fuel combustion	5,322
Electric utilities	445
Industry	1,178
Residential wood burning	3,300
Other	399
Industrial processes	7,590
Chemical and allied product manufacturing	1,081
Metals processing	1,678
Petroleum and related industries	366
Waste disposal and recycling	3,792
Other industrial processes	599
Onroad vehicles	49,989
Light-duty gas vehicles and motorcycles	27,382
Light-duty gas trucks	16,115
Heavy-duty gas vehicles	4,262
Diesels	2,230
Nonroad engines and vehicles	25,162
Recreational	3,616
Lawn and garden	11,116
Aircraft	1,002
Light commercial	4,259
Other	5,169
Miscellaneous	9,378
Slash or prescribed burning	6,152
Forest wildfires	2,638
Other	588
Total	97,441

Source: EPA 2001a.

that the mixture of air and fuel was adequate to provide vehicle performance, but the carburetor could never be truly optimized for emissions control across the entire operating range of the engine. Carbureted engines featured an automatic choke that increased the fueling rate during cold start and initial operation. Depending on choke calibration, CO and HC emissions could be substantial during cold starts. This type of nonfeedback fuel metering is termed an open-loop system.

Since the middle 1980s, modern computer-controlled engines have used electronic fuel injectors rather than carburetors to deliver fuel to cylinders in automobiles and most light-duty trucks. Using closed-loop control, the engine computer system reads the signal from an O_2 sensor in the exhaust system and adjusts the air-to-fuel ratio to help maintain stoichiometric combustion. That feedback provides just enough air to combust the fuel but maintains the maximal catalytic-converter efficiency for control of CO, HC, and nitrogen oxides (NO_x).

During hard acceleration and high-speed operations, however, engine computers often use fuel-enrichment strategies to enhance engine performance for short periods and to protect sensitive engine components from high-temperature damage. Likewise, fuel-enrichment strategies are often used during cold starts (as discussed later). Thus, in modern engines, CO emissions are prominent primarily during enrichment associated with heavy loads, hard accelerations, and cold starts.

The onboard diagnostics system installed in model year 1996 and newer vehicles, known as the OBDII system, can help to detect problems during vehicle operations that increase CO emissions. The OBDII system uses sensors to monitor and modify the performance of the engine and emissions-control components. The onboard computer detects signals from the sensors to identify sensor and control-system failures, illuminating the malfunction indicator light on the vehicle dashboard and storing the fault codes (know as diagnostic trouble codes) for later analysis. In a garage setting, mechanics can download the OBDII fault codes from the onboard computer with a diagnostic analyzer ("scan tool"). The codes identify emissions-control systems and components that are malfunctioning. However, some components of OBDII systems (such as exhaust-gas recirculation and O_2 sensors) are often disabled by the engine computer under conditions in which the manufacturer cannot guarantee the components' performance (J. Cabaniss, Association of International Automobile Manufacturers, personal communication, July 10, 2001). That tends to be the case for vehicles operating at temperatures below 20°F. It may be prudent to take steps to ensure that manufacturers certify their OBDII systems to lower operating temperatures than currently required. This is especially true if many northern locations begin adopting OBDII inspection and maintenance (I/M) systems, where the OBDII system is relied upon to determine whether a vehicle fails or passes an emissions test. When a significant number of sensors become inoperative, the OBDII system will also have less ability to alert vehicle owners of potential emissions-system failures.

Cold-Start Emissions

Under cold-start conditions, the engine computer commands the fuel injectors to add excess fuel to the intake air to ensure that enough fuel evaporates to yield a flammable mixture in the engine cylinders. A typical engine-computer strategy injects several times the stoichiometric amount of fuel during the first few engine revolutions, using a fixed fueling schedule to reach idling conditions. Excess fuel continues to be injected until the engine and O_2 sensor are warmed up and the exhaust-catalyst inlet temperature reaches about 250-300°C (482-572°F), sufficient for the catalyst to oxidize CO to CO_2. This open-loop operation, before the catalyst reaches peak efficiency, can continue for several minutes at low ambient temperatures. It is responsible for most of the emissions of CO, air toxics, and unburned HCs from properly operating modern vehicles. Once the engine and emissions-control systems are warmed up, combustion becomes stoichiometric, and CO is converted to CO_2 in the catalyst, keeping CO emissions very low under typical operating conditions. Typical warmup times under mild ambient conditions, around 70-80°F, can be about 1 minute (min) for modern catalysts and even as short as a few seconds for modern close-coupled catalysts (catalysts close to the engine). When ambient temperatures are -20°F or lower, however, catalyst and engine warmup times can exceed 5 min (Sierra Research 1999).

The amount of fuel enrichment required for a cold start of the engine is a function of engine design and ambient air and coolant temperatures. As the temperature in the combustion chamber gets lower, gasoline vapor pressure decreases, and additional fuel is required to ensure ignition. Gasoline sold during the winter in cold climates has a higher vapor pressure than gasoline sold in the summer to assist cold-start ignition. In summer, the vapor pressure is reduced to minimize evaporative emissions.

Below about 0°F, engine starting becomes more difficult, so external means, such as adjusting the fuel vapor pressure, must be devised to ensure that enough fuel vaporizes. A popular method for assisting engine startup is to use engine-block heaters, which heat the engine coolant to allow easier cold starting. These plug-in heaters reduce the time for engine and catalyst warmup under cold conditions and so help to avoid the extremely high enrichments used by fuel-management control units at low temperatures. Use of plug-in heaters in very cold conditions is also an effective method of reducing CO emissions (Sierra Research 1999).

New-Vehicle Certification Programs

Emissions of CO during cold-start conditions and hot stabilized operations are most efficiently controlled through engineering design, that is, by the original equipment manufacturers, in new vehicles. The most important and effective vehicle emissions-control strategy for CO has been the nationwide introduction in 1980 and 1981 of more stringent vehicle certification requirements designed to reduce exhaust emissions of CO, HCs, and NO_x. In 1981, EPA introduced emissions certification standards for passenger cars that provided a 3.4-g/mile (mi) limit for CO. That limit is still in place. The new-vehicle certification program appears to have contributed to reductions of about 36% in CO emissions from mobile sources from 1980 to 1999 (EPA 2001a).

Though the emissions certification standards for CO for passenger cars have not changed, there have been myriad regulations resulting in reductions in CO from the vehicle fleet. HC emissions were greatly reduced by Tier 1 vehicle emissions certification standards (introduced in model year 1994) and the National Low Emissions Vehicle (NLEV) standards (introduced in model year 1999). Because CO is readily reduced by the same vehicle emissions-control technology that reduces HC, more stringent HC standards have resulted in concomitant reductions in CO. The Tier 1 standards also required emissions-control system durability to increase to 100,000 mi (from 50,000 mi), which improved the robustness of emissions-control systems and tightened CO standards for some light-duty trucks. Tier 2 vehicle emissions certification standards, which will be introduced in model year 2004, will require that all light-duty trucks meet the CO standards for passenger cars. In addition, the Tier 2 standards also greatly reduce the sulfur content of gasoline. As discussed in Chapter 2, reducing fuel sulfur content improves catalyst performance, thereby reducing CO emissions. Thus, the Tier 1, NLEV, and Tier 2 vehicle-emissions and fuel standards have provided and will continue to provide improvements to vehicle CO emissions performance as the fleet turns over.

Since 1994, new cars have been required to meet a CO limit of 10 g/mi on the same FTP cycle conducted at 20°F (40 CFR Subpart C 86.201-94, pages 544-554). However, it is uncertain whether these vehicles achieve the same low emissions at substantially lower operating temperatures. The lower-temperature cold-start test represents only the upper boundary of the temperature range of concern in Fairbanks. Furthermore, some emissions-control-system sensors and actuators are inaccurate at very low temperatures, and the engine computer might ignore their input signals until the engine warms up (J. Cabaniss, Association of International Automobile Manufacturers, personal communication, July 10, 2001). Additional studies on the emissions behavior of vehicles at temperatures below 20°F should be conducted to establish the effect of these extreme temperatures on emissions and control devices. Studies on after-market technologies that could provide benefits similar to those that would be achieved with an enhanced cold-start test should also be undertaken. It should be noted that EPA does not perform in-use vehicle testing to confirm that vehicles certified to the low-temperature standards are meeting those standards.

Reducing the 10 g/mi limit for the 20°F cold-start test or reducing the test temperature may provide additional CO emissions reductions for cold northern regions, such as Fairbanks. Indeed, CAAA90 mandated that, if six or more areas were designated as nonattainment as of July 1, 1997, EPA require cars to meet a cold-start emissions limit of 3.4 g/mi. EPA has yet to make a formal determination of the number of nonattainment areas as of July 1, 1997. A question to be addressed is whether requiring all automobiles sold nationwide to meet a more stringent cold-start test may be an ineffective use of resources, in that it would impose additional emissions controls on vehicles that might never be used in cold climates.

Strategies to Address Emissions-Control Failures

Although progress has been made in controlling CO emissions from vehicles, some problems remain. One concern is that failures in the operation of emission-control systems typically default to fuel-rich conditions that produce higher CO emissions while allowing engine performance to be maintained. For example, when O_2 or temperature sensors are defective, engine computers may default to fuel-rich conditions. Similarly, defective fuel injectors may result in higher CO emissions. A weak spark ignition can cause hard starting, misfires, and poor performance and result in increased CO emissions. Other common failures in the emissions-control system that affect CO emissions are O_2-sensor deterioration, air-injection system defects, and catalyst deterioration. Because engine failures are unavoidable and many such failures cause higher CO emissions, a strategy is needed to identify vehicles with unacceptably high emissions.

Many urban areas use inspection and maintenance (I/M) programs to identify the highest-emitting vehicles and require that they be repaired. High emitters exhibit increased emissions under almost all onroad operating conditions because of failure in emissions-control or fuel-control systems. Not surprisingly, cold-start emissions from high-emitting vehicles are typically also substantially increased. Thus, identification of vehicles as high emitters during the summer months and their repair before winter can yield significant reductions in cold-start emissions. However, some emissions-control system failures may go undetected during an I/M test of a vehicle's tailpipe emissions under idle conditions. Placing a vehicle under a driving load, such as by testing it on a dynamometer, as is done in the IM240 test, has a higher probability of revealing an O_2-sensor malfunction and other defects. Remote-sensing systems, which determine CO emissions by measuring absorption of infrared radiation from a beam directed across the roadway, can also be used to identify high emitters (NRC 2001).

AIR QUALITY MODELS

Air quality modeling is an essential element of air quality management. Models can be used to demonstrate attainment of the NAAQS, evaluate the effects of new construction projects, and conduct further research into what causes pollution episodes and how to predict them. A number of modeling techniques—requiring various levels of scientific expertise, input data, and computing resources—are available for those purposes. The simplest models assume a direct correlation between emissions and ambient pollutant concentrations; the most complicated models resolve temporal and spatial variations in pollutant concentrations and include the effects of meteorology, emissions, chemistry, and topography. Models are also characterized by the size of the problem they address: microscale models simulate pollution from an intersection or point source, mesoscale models simulate metropolitan or multistate pollution, and large-scale models simulate continental or global pollution.

In the attainment demonstration presented in their SIPs, states are required by EPA to model how emissions reductions will lead to the desired air quality improvements. Three types of models have been used to demonstrate attainment of the CO NAAQS: statistical rollback, Gaussian dispersion, and numerical predictive models. The simplest is a statistical rollback model in which the needed reduction in emissions is assumed to be proportional to the required reduction in ambient CO concentrations (ADEC 2001a):

$$\% reduction = \frac{CO_{baseyear} - CO_{NAAQS}}{CO_{baseyear} - CO_{background}},$$

where

$CO_{baseyear}$ = the second highest 8-h average in the base year;
CO_{NAAQS} = the NAAQS of 9 ppm (or sometimes 9.4 ppm); and
$CO_{background}$ = an average regional background CO in the absence of emissions.

Because no information is needed about meteorology or the spatial distribution of emissions in a nonattainment area, EPA has allowed states to use rollback models to demonstrate attainment in smaller cities, rather than the more resource-intensive dispersion and urban-airshed models described below. Although easy to implement, rollback models do not explicitly consider the role of meteorology or the spatial heterogeneity of CO emissions and concentrations.

A second type of model that has been used for CO-attainment demonstrations is a Gaussian dispersion model, which is typically used to simulate CO concentrations for microscale analysis in the vicinity of intersections or along major traffic corridors (EPA 1992). These models simulate how a pollutant is dispersed into the immediately surrounding atmosphere. They assume that the atmospheric concentration of the pollutant is proportional to its emissions and inversely proportional to windspeed and that the resulting spatial distribution of the pollutant is Gaussian (Wayson 1999). Inputs for dispersion models include meteorological data, such as windspeed and inversion strength in the vicinity of the pollutant source, and temporally resolved emissions. Because predicted concentrations are directly proportional to emissions, the accuracy of emissions measurements is crucial to the modeling process. In the case of modeling of intersections, the emissions inventory may be derived from information about traffic patterns, mean speeds, and vehicle-fleet composition. Larger cities have also used Gaussian dispersion models to evaluate the air quality effects of increasing road capacity or other large construction projects (EPA 1992).

Box models, another tool available for microscale analysis of air pollution, have not been used extensively for SIP attainment demonstrations. The "box" is some volume of air into which emissions are injected and in which chemical transformation may take place and air is exchanged with the environment. The air in the box is assumed to be well mixed, so spatial variations in emissions or pollutant concentrations on scales smaller than the box model are not resolved. Box models are particularly useful for understanding how various emissions scenarios and meteorological conditions affect pollutant concentrations. For example, a box model for CO in Anchorage, Alaska, has been used to quantify how mechanical turbulence from roadway traffic might increase the mixing height and reduce CO concentrations on severe-stagnation days (Morris 2001). Limitations of the box-model approach include an inability to include spatial variations and a dependence on assumptions to represent meteorological parameters. Chapter 2 presents a simple box model developed for the Fairbanks nonattainment area.

The most complicated models used for attainment demonstrations simulate how a pollutant concentration varies with time and space over an entire urban area. These numerical predictive models, generally intended for mesoscale analysis, can simulate emissions from multiple sources and the dispersion, advection, and photochemical reactions of gaseous pollutants in the atmosphere. Numerical predictive models, such as the Urban Airshed Model (UAM), have been used for many years to simulate O_3, which is an areawide or mesoscale pollutant. The UAM has also been adapted to simulate CO in Denver, Colorado (Colorado Department of Public Health and Environment 2000). Because of the local nature of high-CO episodes, extensive modeling of the entire urban airshed may be unnecessary for CO-attainment demonstrations. Airshed modeling is resource-intensive, requiring detailed knowledge of an area's meteorology (usually based on the output of a mesoscale weather model constrained by observations), spatially and temporally resolved emissions inventories, and measurements of the pollutant at several locations to allow model evaluation. Highly trained personnel are needed to conduct the simulations. However, a simplified approach of this method may be appropriate in some cases.

More complicated models are not always appropriate for attainment demonstrations, but they can be valuable in improving our understanding of the interactions among atmospheric processes. Even better research tools than the numerical predictive models describe above (such as the UAM) are process numerical models, which allow coupling between processes specific to air quality modeling and meteorology. Process numerical models typically are formulated by adding pollutant emissions, chemistry, and transport into an existing meteorological model rather than simply using the meteorological data as a model input. The relatively nonreactive behavior of CO makes it an ideal chemical species for simulation in a weather model. Predictions of CO, for example, can be straightforward in the National Weather Service Eta Model, which has a horizontal grid framework of 12 × 12 km over the contiguous United States.

Despite advances in air quality modeling capabilities over the last 30 y, many improvements are still possible and needed, particularly in the numerical predictive models, which are used more widely than process numerical models. One problem is that the vertical and horizontal resolution of both types of models is too coarse to capture the variability in pollutant concentrations, which is necessary to identify local hotspots. Most numerical predictive and process numerical models are based on statistical representations of atmospheric motion on scales smaller than the spatial resolution of the models. When unusual meteorological conditions occur, the validity of these representations becomes questionable and could lead to errors in the prediction (Pielke 2002). Models used for regulatory purposes can suffer the loss of realism as a result of such shortcomings.

Various models have been applied to predict future pollutant concentrations, particularly with the goal of identifying conditions that might create an episode. Numerical predictive models can be used, as can simpler empirical models, which attempt to identify statistically significant relationships between specific air quality variables and a set of predictors. Empirical models typically use regression or neural-network techniques to develop a relationship based on observations of meteorological variables and pollutant concentrations. Future air quality can be predicted by using the output of weather-forecast models as values for the predictors. Meteorological forecasts are disseminated daily by the National Weather Service (Hooke and Pielke Jr. 2000), but only a few areas in the country provide short-term air quality forecasts, usually only for O_3, which is an areawide pollutant.

SUMMARY

CO is a pollutant that impairs the ability of blood to carry O_2 to body tissues. Exposure to CO at sufficiently high concentrations can cause headaches, exacerbate heart problems, lengthen reaction times, and affect fetal development. On the basis of a compilation of scientific knowledge about the relationship between various concentrations of ambient CO and their adverse health effects, EPA set CO standards of a maximum 1-h average concentration of 35 ppm and a maximum 8-h average concentration of 9 ppm. Even though the standards are for outdoor air, they help to improve indoor air. CO remains in the outdoor air long enough to penetrate the indoor environment and is not removed by filtration. Because the measures taken to reduce CO emissions typically result in reduced emissions of copollutants, such as $PM_{2.5}$ and air toxics, the health benefits of CO reductions can be greater than those associated with reduced CO concentrations alone.

Controls on CO emissions, particularly in the form of improved vehicle technology, have led to significant reductions in ambient CO concentrations throughout the United States. However, a few locations still experience concentrations that approach or exceed the CO 8-h health standard. Most of those areas have meteorological conditions (such as frequent inversion or stagnation conditions) or topography (such as being situated in a mountain valley) that inhibit ventilation and allow CO to accumulate at high concentrations near the surface. One such location is Fairbanks, Alaska. The task of the Committee on Carbon Monoxide Episodes in Meteorological and Topographical Problem Areas was to assess approaches for predicting, assessing, and managing episodes of high CO concentrations in meteorological or topographical problem areas. The committee's charge called for Fairbanks to be the focus of this interim report. The rest of this report describes the CO problem there, including its physical characteristics, the emissions-control strategies used there, and the prospects for the area to remain in attainment with the NAAQS for CO.

2

Fairbanks Case Study

The committee was charged to develop this interim report with a focus on Fairbanks, Alaska, one of the most challenging carbon monoxide (CO) nonattainment areas in the country. The Fairbanks North Star Borough stands out as the only serious CO nonattainment area with a population under 100,000 and little industry. In Fairbanks, vehicle emissions, meteorology, and topography combine to produce conditions conducive to high ambient CO concentrations.[1] First, the low winter temperatures increase CO emissions from passenger cars, vans, sport utility vehicles, and light-duty trucks, especially during cold starts. Second, Fairbanks experiences strong and long-lasting ground-level temperature inversions during the winter, trapping pollutants close to the ground. Third, Fairbanks is in a river valley, which exacerbates the strength of inversions and decreases windspeeds, further reducing pollutant dispersion.

As discussed in later sections, it is not the coldest days that have the highest CO concentrations. At very low temperatures (down to -50°F), the use of engine-block heaters, which reduce cold-start emissions, and the presence of

---

[1] In the Fairbanks case, the committee concluded that large stationary sources, such as power plants and refineries, do not contribute substantially to high CO concentrations in the areas around the CO monitors. Thus, the committee did not consider stationary-source controls in this interim report. However, the importance of stationary sources and their controls will be considered as the committee produces its final report.

ice fog (from water vapor that is emitted and freezes), which raises the inversion height, help prevent accumulation of high ambient CO concentrations. The current problem with high ambient CO concentrations tends to occur (five out of the last six exceedances of the 8-h National Ambient Air Quality Standard [NAAQS]) at temperatures of roughly 0-20°F, when ice fog is not present and the use of engine-block heaters is not necessary for starting.

This chapter discusses the CO problem in Fairbanks, beginning with the physical and demographic characteristics of the area; the management strategies used to control the CO problem; and a simple air quality model for coupling meteorology and vehicle emissions. Lessons learned from the Fairbanks case are summarized in the committee's findings and recommendations in the Summary of this interim report and serve as a basis for the development of a final report that will include assessments of other CO problem areas. The specific health effects of exposure to CO in Fairbanks are not discussed, because no human exposure or epidemiological data were available.

GEOGRAPHICAL, METEOROLOGICAL, AND SOCIETAL CONTEXT

Physical Setting

Fairbanks is in a floodplain on the north shore of the Tanana River, just upstream of its confluence with the Chena River. The city is open to the south and southeast, with a very gradual slope in that general direction from the Tanana River up to the foothills of the Alaska Range, some 75 km away. To the west and north, residential areas extend into the Yukon-Tanana uplands a few hundred meters above the city. Level ground extends some 35 km to the east except for Birch Hill northeast of town. Figure 2-1 shows a topographical representation of Fairbanks and the surrounding area.

The meteorology is fairly typical of interior Alaska and other continental high latitudes. In the typical winter pattern, anticyclones (high-pressure systems), weak lows, and other weak pressure fields dominate the weather, all with moderate windspeeds. Strong pressure gradients and cyclonic (low-pressure) systems are unusual in winter. For 1996 through 2001, all exceedances of the 8-h CO health standard in Fairbanks have occurred with southeasterly winds aloft. These winds, which travel over the Alaska Range, are associated with counterclockwise geostrophic flow around a low-pressure system in or near the Gulf of Alaska.

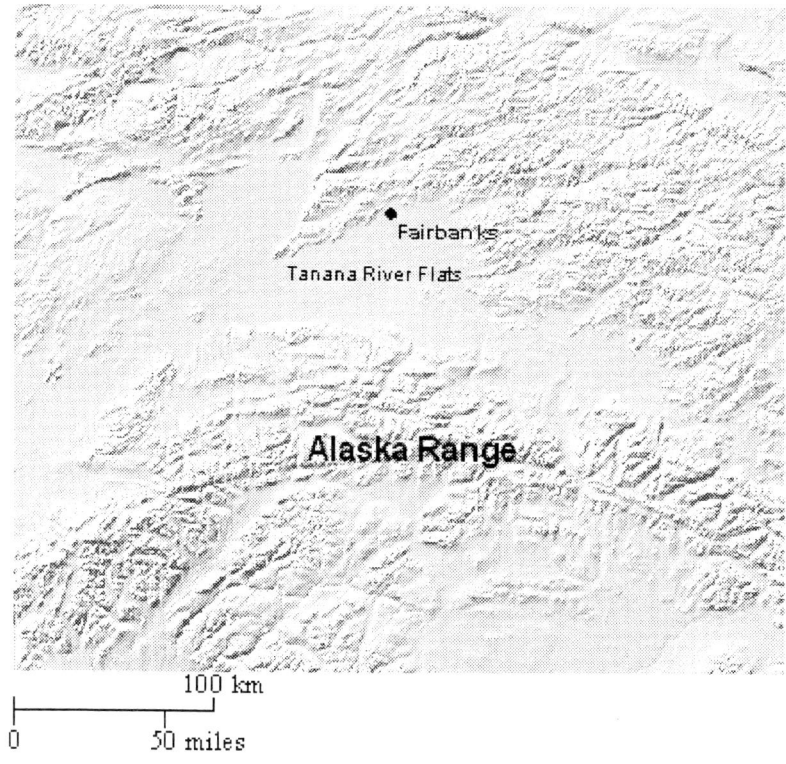

FIGURE 2-1 Map of topography in the Fairbanks area.

Ground-based inversions of considerable strength (typically a few degrees Celsius per 100 m but sometimes much stronger) topped by weaker inversions reaching as high as about 1-2 km are normal in winter and can occur anytime during the year. A surface inversion due to net energy loss from the surface occurs in the few meters closest to the ground, although the weaker inversion topping it may be caused by subsidence or transport of warmer air aloft. The combination of high albedo (reflection of sunlight due to snow cover) and the low solar elevation (failure of the sun to rise high in the sky) characteristic of northern latitudes in winter creates little heating of the ground and weak vertical mixing between the surface and overlying air. With clear skies and low absolute water-vapor content, the ground loses considerably more energy by radiation to space than it is able to absorb from the sun. Those surface conditions may persist in Fairbanks for days, and the situation is exacerbated by the insulation provided by high-albedo snow cover. Although such an inversion may weaken or even dissipate during the middle of the day, it tends to become reestablished or strengthened throughout the late afternoon and into the night. The upper part of the inversion appears to be associated with subsiding (downward) southeasterly flow crossing the Alaska Range. Although the lack of surface warming in winter is common, it now appears that recent exceedances occurred with the upper-level inversion also in place.

Observations in Fairbanks and Anchorage suggest that solar heating of the ground alone is insufficient to produce any substantial mixing layer with solar elevations less than 6° over bare ground or 12-15° over deep snow (Bowling 1985). As shown in Figure 2-2, solar elevations in winter are very low in Fairbanks, even at solar noon.[2] Thus, solar heating seldom breaks the ground inversion from November to February, even without snow cover. Clear skies are much more likely in January through March than in November (Figure 2-3); the low incoming solar heating may help to explain the greater frequency of exceedances during these months.

Inversion measurements (temperature as a function of height) provide an indication of the strength of a ground-level inversion. Ordinary sounding measurements are not sufficient to resolve strong inversions near the ground, although they do at times record inversion strengths larger than 10°C/100 m. In most areas, inversions of several degrees per 100 m would be considered strong. Three limited datasets support such high inversion strengths:

- Traverse data of temperature differences between vehicle-mounted thermistors 0.5 m and 2 m above the ground showed inversions of about 1°C/1.5 m (67°C/100 m) in ice fog and as much as 3°C/1.5 m (200°C/100 m) in clear air (Bowling and Benson 1978). Background inversions in the first 2 m are obviously extremely strong, but nonlinearity in inversion strength makes it difficult to extrapolate this observation to higher altitudes. Such strong low-level ground inversions are unlikely to persist in the downtown area, because heat is normally added with CO in automobile exhaust plumes and mechanical turbulence associated with vehicles can cause further mixing.

- Three tethered-balloon measurements in the downtown area during high-CO conditions indicated a "mixing layer" from the ground up to 6-30 m

---

[2] At 40°N latitude (running from the middle of New Jersey to northern California), the sun is 6° above the horizon about a half-hour after sunrise; in Fairbanks, at 65° N latitude, the sun does not reach this elevation from the middle of November to late January, even at solar noon. The Arctic Circle is at 66.5° N latitude.

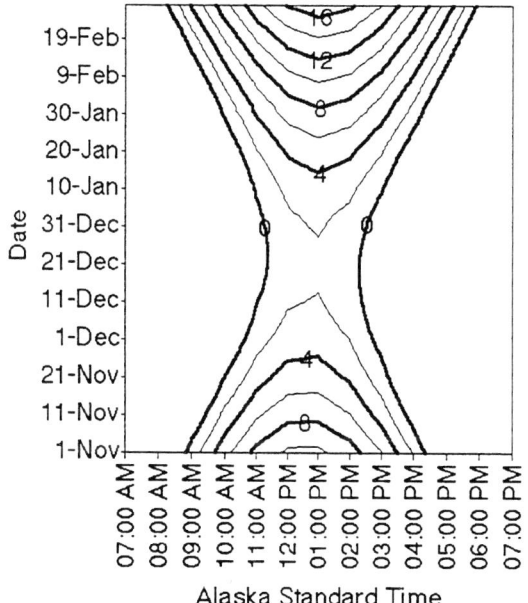

FIGURE 2-2 Solar elevation for Fairbanks in degrees as a function of time of day and day of year.

(Bowling 1986). Background measurements made outside the downtown area with the same tethered balloons showed ground-based inversions of 10-40°C/100 m.

- Borough measurements of temperatures 3, 10, and 23 m above the ground are taken on a meteorological tower in the downtown area (Figure 2-4). Inversion strengths of as much as about 30-40°C/100 m have been measured on the tower during episodes. For example, Figure 2-5 shows the inversion strengths measured on the tower during an exceedance of the CO standard in November 1999. The inversion, measured between 10 and 23 m altitude, averaged 21°C/100 m on November 19, 1999, with maximum hourly values of about 30°C/100 m.

Inversions of the strength observed in Fairbanks cause the buildup of CO in several ways, with stronger inversions associated with higher CO concentrations. First, such inversions strongly inhibit vertical mixing. Second, inversions inhibit momentum transfer, thereby preventing winds aloft from reach-

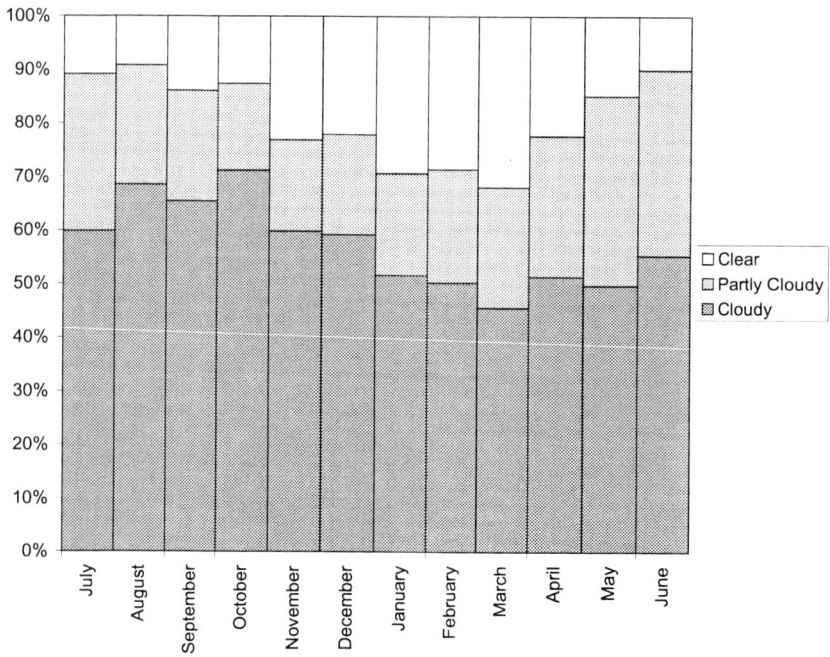

FIGURE 2-3 Fractions of clear (0-2 oktas [eighths of sky]), partly cloudy (3-6 oktas), and cloudy (7-8 oktas) by month, based on NOAA 1961-1990 averages.

ing the surface—a particular problem when warm air advected from the south across the Alaska Range does not reach the ground in Fairbanks. The surface winds that exist in Fairbanks during winter are probably controlled by gravity drainage from higher elevations to the north, often in a direction opposite to that of winds just a few hundred meters higher. These surface winds are difficult to characterize in the downtown area; 90% of the time, winter windspeeds are at or below the minimal starting speed (3 mph) of a standard anemometer. Finally, the combination of strong ground-level inversions with surface temperatures and winds that probably vary widely over horizontal distances of 0.1 km suggests that inversion strength (and probably mixing height) may be highly variable spatially (Holmgren et al. 1975).

February has been the month with the most exceedances over the last 5 y. In February, the reestablishment (or strengthening) of ground inversions occurs rapidly near 5:00 p.m., trapping pollutants emitted when sunset and maximal traffic occur near the same time (Bowling 1984, 1986). The reestablish-

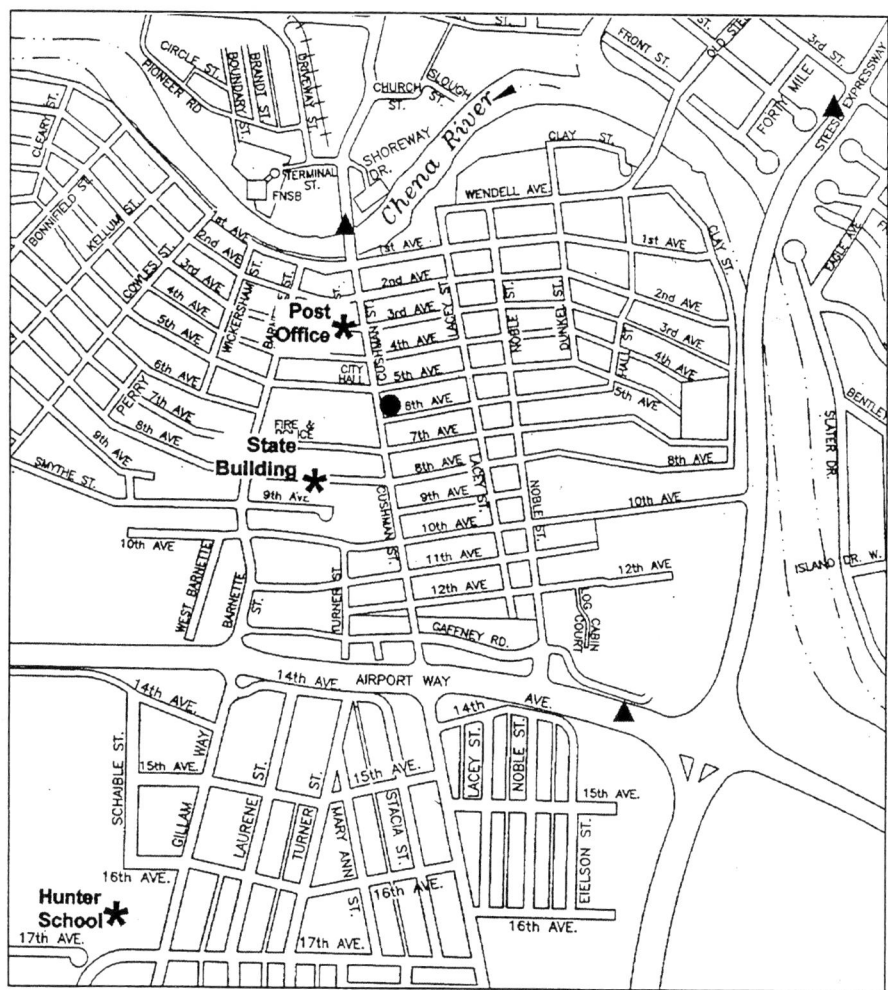

FIGURE 2-4 Map of downtown Fairbanks. Asterisks indicate CO monitoring sites; triangles indicate where vehicles are counted. The meteorological tower is indicated by a dot.

ment of the ground-level inversion in late afternoon and early evening is clearly visible in the inversion-strength measurements made from the meteorological tower in downtown Fairbanks (S.A. Bowling, Fairbanks Meteorology, in preparation, 2002).

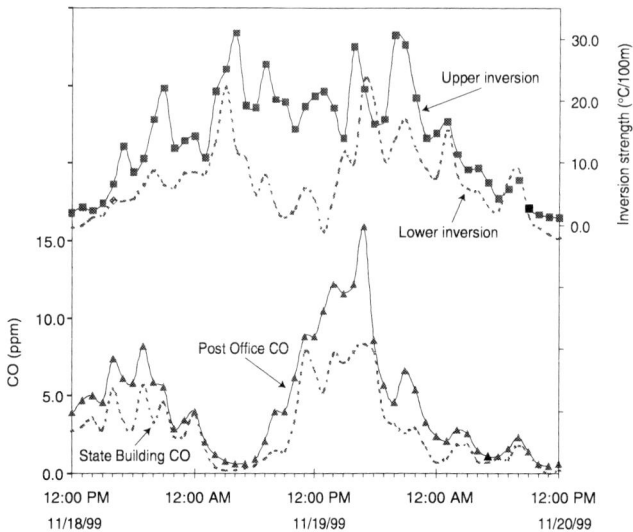

FIGURE 2-5 CO concentrations (ppm) and inversion strengths (degrees Celsius/100 m) measured during an exceedance of the CO standard on November 19, 1999. The inversion strengths were calculated from temperature measurements taken on the meteorological tower. The upper inversion was measured between 10 m and 23 m altitude and the lower inversion was measured between 3 m and 10 m.

Demographics

As pointed out earlier, Fairbanks is a small city to have such a severe air pollution problem. Its 2000 population of only 30,224 was spread over an area of 32.8 mi^2. The entire Fairbanks North Star Borough had a population of 83,632 in an area of 7,361 mi^2, growing from 77,720 in 1990. Over the last decade, the city population has been nearly constant, most of the growth being in the outlying parts of the borough.

The borough's population can be greatly affected by large construction projects and government spending. For example, the construction of the Trans-Alaskan Oil Pipeline caused the borough's population to rise from 45,864 in 1970 to 72,037 in 1976. After completion of the pipeline, the population fell to 51,659 in 1981. Spending of state oil revenue by local and state government helped the borough's population rebound to 75,079 in 1985.

The economy of the borough depends heavily on federal, state, and local government sectors, which combine to account for 30% of jobs. Some 20%

of the borough's population are military or military dependents. Retail trade and services account for 18% and 26%, respectively, reflecting the importance of Fairbanks as a tourist center and service center for most of the state north of the Alaska Range.

The population is relatively young. In 2000, the median age of the borough's residents was 29.5 y, compared with 32.4 y in the state and 35.3 y in the United States as a whole. Only 9.5% of households had members 65 y old or older, compared with 11.9% in the state and 23.4% in the United States as a whole. Male residents made up 52.2% of the population in 2000, compared with 51.7% in the state and 49.1% in the United States (U.S. Census Bureau 2000).

Local transportation is largely by motor vehicle. Traffic counts show that vehicle-miles traveled (VMT) in the area increased from an estimated average of 665,398 mi/day (d) in 1990 to 752,992 mi/d in 2000 (ADEC 2001a). The VMT growth rate exceeds the growth in population of the borough, indicating that residents are driving more on a per capita basis, a trend observed throughout the United States. The average hourly traffic counts observed at three locations in Fairbanks (see Figure 2-4) are shown in Table 2-1 for the CO seasons of 1995-1996 through 2000-2001. Traffic has increased by 6-7% on the Steese Expressway, decreased by 3% on Airport Way, and remained fairly constant at Cushman Street and the Chena River in downtown Fairbanks. Increasing residential, commercial, and retail development outside the city of Fairbanks probably contributes to the increase in VMT and in traffic on the Steese Expressway; however, little additional data are available to confirm this hypothesis. Changing traffic patterns in the city will change the spatial distribution of emissions and ambient CO.

CO DATA

The borough has measured CO in downtown Fairbanks since 1972. The locations of the three monitoring sites currently in operation are shown in Figure 2-4. Although the sites were run year-round in the past, they are now operated only from October 1 through March 31 (ADEC 2001a). The measurements are made with commercially available monitors that have an uncertainty on the order of 0.1 ppm and a detection limit of around 0.1 ppm. Hourly average concentrations are recorded. The borough provided these hourly data to the committee for the CO seasons (November through

TABLE 2-1 Average Hourly Traffic Counts for Fairbanks, Alaska, During Winter Months

	Airport Way	Steese Expressway	Cushman St. and Chena River
Nov. 1995 – Feb. 1996	644	758	557
Nov. 1996 – Feb. 1997	641	747	558
Nov. 1997 – Feb. 1998	648	778	568
Nov. 1998 – Feb. 1999	636	779	536
Nov. 1999 – Feb. 2000	615	786	539
Nov. 2000 – Feb. 2001	626	814	575
% change 95-96 to 00-01	-2.75	7.31	3.16
% change 95-97 to 99-01[a]	-3.35	6.27	-0.18

[a]Referring to the average of the last two seasons (Nov. 1999-Feb. 2000 and Nov. 2000-Feb. 2001) compared with the first two seasons (Nov. 1995-Feb. 1996 and Nov. 1996-Feb. 1997), reducing the possible effect of one unusual year.

February) of 1995-1996 through 2000-2001. As described below, a detailed trend analysis of the running 8-h average concentrations computed from these data (a total of 47,980 data points) was conducted by the committee to illustrate statistical techniques for assessing progress in reducing CO concentrations.

Although the three monitoring sites satisfy EPA's monitoring guidelines, they are located quite close together and therefore do not provide much information about the spatial distribution of CO concentrations across the borough. The following analysis of observations at these three sites is limited in that respect, and similar analyses should be conducted in the future with data that are more spatially representative. The Alaska Department of Environmental Conservation (ADEC) has conducted two saturation studies—during the winters of 1999-2000 and 2000-2001—to begin improving the characterization of CO on the regional scale (Guay 2001).

Exceedances in Fairbanks

As in most areas, the number of exceedances of the 8-h NAAQS for CO in Fairbanks has decreased over the last 25 y (Figure 2-6).[3] During the early years of regular CO monitoring in Fairbanks, there were more than 130 d with exceedances per year, distributed over all months. The numbers dropped rapidly until 1987-1988 after which they decreased more slowly. Exceedances since then have occurred entirely during the winter months of November through March. As of January 2002, there have been no exceedances since February 2000.

As shown in Figures 2-7 and 2-8, exceedances vary by day of the week and hour of the day. Figure 2-7 shows that substantially fewer exceedances occur on weekend days; this reflects the decreased use of vehicles on these days and possibly the timing of vehicle operations. Figure 2-8 shows the number of monitor hours[4] with 1-h CO concentrations over 9 ppm for the six most recent exceedance days for each hour of the day. This figure shows that high CO concentrations have recently been most likely around 5:00-6:00 p.m. Five of the six most recent exceedance days have been in February. As described in the previous section, strong temperature inversions tend to occur around sundown when people are leaving work and starting their cars.

Temperatures on seven of the eight most recent exceedance days ranged from -4.2°F to 28.6°F. The potential importance of exceedances occurring in the "relative warmth" of about 0-20°F is that, at this temperature range, vehicles do not usually need to be plugged in to start easily. It is thought that the high emissions from cold starts of vehicles that had not been plugged in contributed to the exceedances.

CO exceedances in the borough are a function of location. Of the last six exceedances, four had the highest measured CO concentrations at the Post Office monitor (Courthouse Square), one at the Hunter School, and one at the State Office Building. Studies by the ADEC Air Monitoring Section, which deployed temporary monitors over a wider area, indicate that CO concentra-

---

[3]The year here is defined to cover the winter season and is the period from July 1 of one calendar year to June 30 of the following year. A violation of the NAAQS occurs when the standard is exceeded for the second time and all subsequent times during such a year.

[4]If the 1-h average CO concentration was over 9.0 ppm at two of the three monitors in downtown Fairbanks at a particular hour, that would count as two monitor hours.

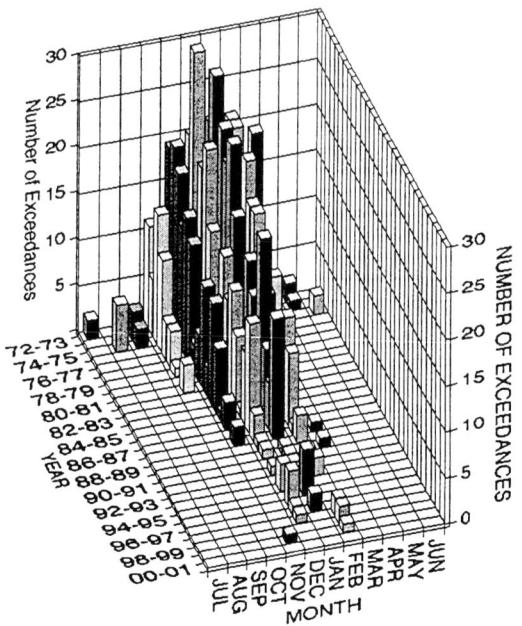

FIGURE 2-6 Number of days with exceedances of the 8-h CO standard per month per year.

tions vary widely throughout the borough in both horizontal and vertical dimensions. Variability in CO as a function of altitude may reflect heterogeneity in the mixing-layer height. However, it is not known how large the area of high CO is or generally how CO concentrations vary spatially in the nonattainment area.

BOX 2-1 Recommendations: Meteorological Conditions of Primary Concern

Air quality management in Fairbanks should focus on the 0° to 20°F temperature range. Emissions inventories should be refined and verified and control programs evaluated for their effectiveness with emphasis on that temperature range. In addition, air quality modeling should be developed, conducted, and evaluated for the extreme conditions found in Fairbanks in winter.

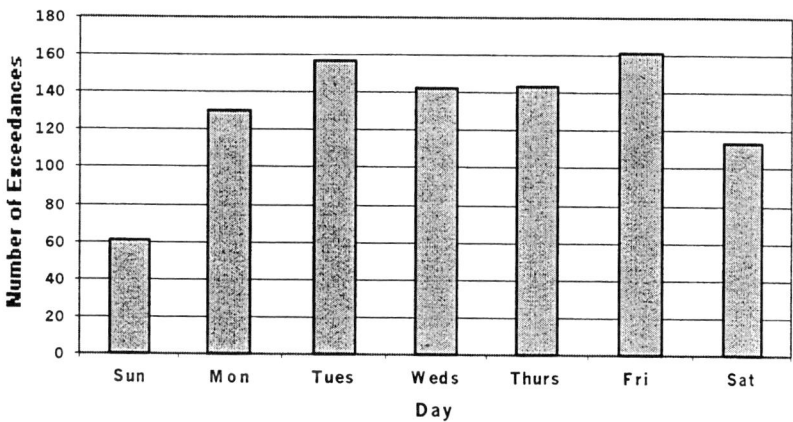

FIGURE 2-7 Number of borough exceedances of the 8-h NAAQS for CO for the period 1972-2001, by day of the week. Source: Data provided by Paul Rossow, Fairbanks North Star Borough.

Recent CO Trends

Concentrations of CO measured at the three monitoring stations in downtown Fairbanks during November through February (the CO season) in 1995-1996 through 2000-2001 were analyzed by the committee.[5] The results of the analyses are shown in Table 2-2. The first maximum, the second nonoverlapping maximum, and several percentiles (the 75th, 90th, 95th, and 99th) were examined as summary measures of the distribution of the running 8-h average CO concentrations observed each season. The percentiles were analyzed because they are more statistically robust benchmarks for measuring progress in reducing CO than are extreme values, like the second nonoverlapping maximum. Extreme values are usually highly variable, especially in the presence of extreme meteorological conditions such as those in Fairbanks. At the same time, an analysis of the central tendency statistics is also not appropriate, because for CO, like many other environmental

---

[5]A dataset containing 1-h average CO concentrations was kindly provided by Paul Rossow, the borough air quailty specialist. Ms. Susan Alber, Department of Biostatistics, UCLA, assisted with the analyses reported in this section.

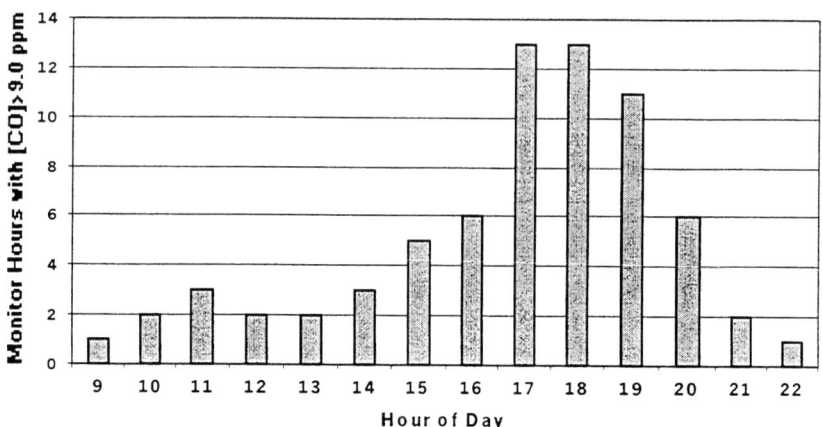

FIGURE 2-8 The number of monitor hours with CO concentrations >9.0 ppm during the most recent exceedance days of 2/23/98, 2/24/98, 2/11/99, 2/16/99, 11/19/99, and 2/8/00.

pollutants, our concern lies with trends in high concentrations and the conditions that produce them.

Across all three monitoring stations, the summary measures decrease over the 6-y period. For each summary measure and each monitoring station, the decreasing trend was summarized into an annual change rate, defined as the slope coefficient for the linear regression of the six values for the annual summary measure on time (see Table 2-2). Similar linear-regression equations were fitted for each of the six summary measures at each of the three monitoring stations (a total of 18 regression equations).[6] Averaged across the three

---

[6]Further analysis was conducted using quadratic regression equations to examine the goodness-of-fit for the linear-regression equations. The quadratic term for nonlinearity is statistically significant in only one of the 18 equations. For the 99th percentile at the Post Office, the decreasing trend appears to be accelerating in recent years, and the quadratic regression exhibits a modest negative curvature. Nevertheless, the p value for the quadratic term in this regression equation is only 0.03, not a compelling piece of evidence against the linear-regression equation in light of the multiple statistical tests (a total of 18) conducted simultaneously. Therefore, it appears that the linear-regression equation is the appropriate specification for the trend analyses.

TABLE 2-2 Trends in CO (in ppm) Measured November Through February at Fairbanks Monitoring Locations

Site	Year	75th Percentile	90th Percentile	95th Percentile	99th Percentile	Second Nonoverlapping Maximum	First Maximum
Hunter School	1995-1996	2.68	4.16	5.08	7.29	8.84	11.56
Hunter School	1996-1997	3.25	4.86	6.15	8.24	9.81	12.75
Hunter School	1997-1998	2.68	4.30	5.41	7.45	10.01	10.44
Hunter School	1998-1999	2.55	3.70	4.75	6.95	9.75	9.88
Hunter School	1999-2000	1.94	2.99	3.93	5.79	6.76	8.60
Hunter School	2000-2001	2.22	3.24	3.91	5.12	6.93	8.34
Annual change rate[a]		-0.18	-0.31	-0.38	-0.53	-0.54	-0.83
T-test for annual change rate		-2.31	-2.93	-2.73	-3.52	-1.87	-4.50
Post Office	1995-1996	3.28	4.93	6.31	9.03	11.15	15.16
Post Office	1996-1997	3.66	5.31	6.50	8.45	9.76	13.26
Post Office	1997-1998	2.68	4.10	5.15	8.36	10.35	12.13
Post Office	1998-1999	3.03	4.31	5.20	7.58	9.43	10.26
Post Office	1999-2000	2.64	3.80	4.78	6.54	9.68	11.48
Post Office	2000-2001	2.28	3.45	4.16	5.40	7.01	8.57
Annual change rate[a]		-0.22	-0.33	-0.45	-0.70	-0.62	-1.15
T-test for annual change rate		-2.94	-4.08	-5.65	-7.52	-3.08	-5.22
State Building	1995-1996	2.59	3.93	4.90	7.81	9.23	13.09

(Continued)

TABLE 2-2 *Continued*

Site	Year	75th Percentile	90th Percentile	95th Percentile	99th Percentile	Second Nonoverlapping Maximum	First Maximum
State Building	1996-1997	3.16	4.55	5.71	7.60	9.14	12.20
State Building	1997-1998	2.53	3.78	4.71	6.68	9.20	10.76
State Building	1998-1999	2.40	3.58	4.25	6.58	7.36	9.09
State Building	1999-2000	1.90	2.94	3.69	6.11	8.78	9.72
State Building	2000-2001	2.03	2.99	3.69	4.91	6.34	8.14
Annual change rate[a]		-0.19	-0.28	-0.36	-0.55	-0.50	-0.97
T-test for annual change rate		-2.63	-3.28	-3.38	-7.48	-2.39	-6.73

[a]Annual change rate is defined as the regression slope coefficient for rate of change in concentration.
Note: CO observations are 8-h running averages of hourly average date. Approximately 2,880 data points are available at each site for each year.
Source: Data provided by Paul Rossow, Fairbanks North Star Borough.

monitoring stations, the first maximum decreased by about 1 ppm each year; the second nonoverlapping maximum and the 99th percentile decreased by about 0.5 ppm each year; the 95th percentile, 90th percentile, and 75th percentiles decreased by about 0.4, 0.3, and 0.2 ppm, respectively, each year. The t-test statistics for the slope coefficients indicate that the decreases are statistically significant at the conventional 5% level for 17 of the 18 annual change rates.

As shown in Table 2-3, the ratio of the annual change rate to the value of the summary measure in 1995-1996 is about 7% for all the measures at all three sites. That implies that for the 75th percentile and higher, CO concentrations have been declining by about 7% each year in the Fairbanks nonattainment area over the period 1995-1996 to 2000-2001.

The analysis of the most recent 6 y shown in the current analysis only provides limited representativeness of the trends in CO concentrations in Fairbanks. Trends in ambient CO concentrations in Fairbanks have also been reported by EPA for 1986-1995 (EPA 2000a) and by ADEC for 1972-2000 (ADEC 2001a). EPA (2000a) noted a downward trend of the hourly average CO concentrations for 1986-1995, although the median and interquartile range of the daily maximum 8-h average concentration remained relatively unchanged for 1989-1995. ADEC (2001a) reported a strong downward trend in the number of exceedance days per year and in the highest and second highest CO concentrations observed. However, based on the committee's independent examination of Figure III.C.3-3, it should be noted that the decline reported in ADEC (2001a) and confirmed in the preliminary analysis presented above is not uniform over time. Further analysis beyond the most recent 6 y is therefore warranted.

Air Quality Alerts

The borough has been broadcasting alerts since at least the winter of 1979-1980 on days when analysts predict that there is a good chance that the CO 8-h average NAAQS of 9 ppm will be exceeded. Forecasts of air quality are made daily by the borough's air quality specialist on the basis of observed CO concentrations, meteorological conditions, and expert judgment. Table 2-4 shows the results of forecasting during the winters of 1997-1998 through 2000-2001. Note that the 4 d with the highest observed 8-h average CO concentrations were called correctly. Rows in italics were days on which exceedances oc-

TABLE 2-3 Ratio of Annual Change Rate to Statistic for 1995-1996

Site	75th Percentile	90th Percentile	95th Percentile	99th Percentile	Second Nonoverlapping Maximum	First Maximum	Overall Mean	Overall Standard Deviation
Hunter School	-0.068	-0.074	-0.074	-0.073	-0.061	-0.072	-0.070	0.005
Post Office	-0.067	-0.068	-0.072	-0.078	-0.056	-0.076	-0.069	0.008
State Building	-0.074	-0.071	-0.073	-0.070	-0.054	-0.074	-0.069	0.008
Mean	-0.070	-0.071	-0.073	-0.074	-0.057	-0.074	-0.070	
Standard Deviation	0.004	0.003	0.001	0.004	0.004	0.002	0.007	

TABLE 2-4 Maximum 8-h Average CO (ppm) in Fairbanks for Exceedances and Alerts During the Winter Seasons of 1997-1998 through 2000-2001

Winter	Date	Day	Forecasted CO	Observed CO	Observed CO Post Office	State Building	Hunter School
97-98	19 Dec	Friday	11	**12.1**	**12.1**	10.8	10.0
97-98	20 Jan	Tuesday	9	7.3	7.2	6.5	7.3
97-98	21 Jan	Wednesday	9	2.5	2.5	1.9	2.4
97-98	4 Feb	Wednesday	9	8.1	8.1	6.0	7.7
97-98	5 Feb	Thursday	10	8.7	6.7	6.6	8.7
97-98	6 Feb	Friday	9	5.7	5.2	4.1	5.7
97-98	*23 Feb*	*Monday*	*No Alert*	*10.2*	*10.2*	*8.0*	*7.5*
97-98	**24 Feb**	**Tuesday**	**12**	**11.1**	10.4	**11.1**[a]	10.4
97-98	25 Feb	Wednesday	11	5.7	5.3	3.8	5.7
98-99	12 Jan	Tuesday	9	8.2	8.2	7.1	7.6
98-99	*11 Feb*	*Thursday*	*No Alert*	*9.8*	*9.4*	*7.4*	*9.8*[b]
98-99	*16 Feb*	*Tuesday*	*No Alert*	*10.3*	*10.3*	*9.1*	*9.9*
99-00	**19 Nov**	**Friday**	**9**	**11.2**	**11.2**	7.3	5.8
99-00	5 Feb	Saturday	9	3.3	3.3	1.4	3.0
99-00	**8 Feb**	**Tuesday**	**10**	**11.5**	**11.5**	9.7	8.6
99-00	9 Feb	Wednesday	9	2.8	2.8	1.9	2.2

(Continued)

TABLE 2-4 *Continued*

Winter	Date	Day	Forecasted CO	Observed CO	Observed CO Post Office	Observed CO State Building	Observed CO Hunter School
00-01	22 Nov	Wednesday	8	7.0	7.0	6.3	6.9
00-01	21 Dec	Wednesday	9	8.6	8.6	8.1	8.3
00-01	22 Dec	Thursday	9	6.3	6.3	5.3	4.6

[a]There was a missing 1-h average at 3:00 p.m.
[b]The maximum at Hunter School occured during the 8-h period that ran from 7 p.m. on Feb. 11 to 2 a.m. on Feb. 12.
Note: Rows with bold entries are for correct exceedance calls. Rows with italics are for missed exceedances.

curred but for which there was no alert. During the four seasons, 16 alerts were called; four alerts were correct (an exceedance actually occurred), and 12 were called when no exceedance occurred. During the same period, three exceedances occurred that were not forecast.

Preliminary analysis suggests that such alerts have little effect on vehicle counts. But, they may make the public more aware of the CO problem. In addition to making air quality forecasts available to the media, the borough has recently put them online to make them accessible to the public through the internet.

Empirical Modeling

Further analysis has been conducted using the Fairbanks CO data to investigate the factors that affect CO concentrations. Empirical modeling attempts to discern statistically significant relationships between an outcome variable and various predictor variables. In particular, an empirical model was developed using the 1-h average CO concentration at the Hunter School monitoring site[7] as the dependent variable and the following predictors: average hourly vehicle counts at three locations in Fairbanks and average hourly meteorological measurements made at a 75-ft tower in the downtown area. (See Figure 2-4 for locations of the CO monitors, traffic counters, and meteorological monitors.) The meteorological measurements included lower inversion strength (measured 3-10 m above the ground), upper inversion strength (measured 10-23 m above the ground), and windspeed, temperature, and atmospheric pressure (all measured 10 m above the ground). All data were from the six winter seasons from November 1, 1995, to the end of February 2001—a total of 722 d (17,328 h). Only hours with complete data in all seven variables were included in the analysis; 5.8% were excluded because of incomplete data in one or more variables.

Most of the variables showed near-normal distributions, but the distribution of CO concentrations was right-skewed and followed a lognormal distribution approximately. Accordingly, the variable was transformed to log(CO

---

[7]The Hunter School site is at least 50 m from the nearest through street and may be least likely to be influenced by the passage of high-emitting vehicles.

TABLE 2-6 Preliminary Multiple Linear-Regression Analysis for $Log(CO + 0.1)$ on Standardized Units

	Coefficient	Standard Error[a]	P-value[a]
Traffic	0.179	0.0025	<0.0001
Low inversion	0.096	0.0036	<0.0001
High inversion	0.077	0.0035	<0.0001
Windspeed	-0.161	0.0025	<0.0001
Temperature	-0.042	0.0026	<0.0001
Pressure	0.010	0.0024	0.0001

[a]Not adjusted for autocorrelation; given for reference only.

+ 0.1).[8] Table 2-5 shows the correlation matrix for the seven variables. $Log(CO + 0.1)$ has the strongest correlation with windspeed (negative, as expected) and moderate correlations with inversion strengths and traffic (all positive, as expected); low and high inversion strengths are strongly and positively correlated with one another, as might be expected.

Table 2-6 shows the results of a preliminary multiple linear-regression analysis with standardized units.[9] Traffic stands out as the most important (positive) predictor variable; windspeed (negative) is next, followed by the inversion strengths. The R^2 for the model is 0.44. This preliminary regression analysis does not take into account autocorrelation among hourly CO concentrations; therefore the standard errors and p-values are invalid and given as references only. This preliminary analysis also does not address possible nonlinearity in the relationship between $log(CO + 0.1)$ and the predictor variables or possible interactions among the predictor variables. Furthermore, other variables not included in this analysis, such as wind direction, may also be important predictors.

---

[8]The value 0.1 ppm was added to the observed 1-h CO concentration to allow the logarithmic transformation to be taken when the CO concentration was observed to be zero. The distribution of $log(CO + 0.1)$ is approximately normal.

[9]Each variable was standardized by subtracting the mean and dividing by the standard deviation.

TABLE 2-5 Correlation Matrix for Hunter School $Log(CO + 0.1)$, Traffic Count, and Meteorological Variables

	Traffic	Low Inversion	High Inversion	Windspeed	Temperature	Pressure	$log(CO+0.1)$
Traffic	1.00	-0.27	-0.20	0.09	0.14	-0.01	0.29
Low inversion	-0.27	1.00	0.74	-0.13	0.19	-0.10	0.29
High inversion	-0.20	0.74	1.00	-0.16	0.13	-0.07	0.33
Windspeed	0.09	-0.13	-0.16	1.00	0.25	-0.14	-0.45
Temperature	0.14	0.19	0.13	0.25	1.00	-0.23	-0.08
Pressure	-0.01	-0.10	-0.07	-0.14	-0.23	1.00	0.06
$log(CO+0.1)$	0.29	0.29	0.33	-0.45	-0.08	0.06	1.00

Source: Data provided by Aaron Owens, DuPont Central Research.

> **BOX 2-2** Recommendations: Improving Ambient Monitoring in the Borough
>
> In the short term, the ambient-CO monitoring network in the borough should be expanded to measure concentrations over a wider area. In the longer term, vertical distributions of CO concentrations and the wind field should be characterized to support the development and application of modeling approaches better than those now available.

EMISSIONS AND VEHICLE CHARACTERISTICS

CO Inventory and Forecasted Reductions

The most recent inventory of CO emissions in the borough was estimated by ADEC (ADEC 2001a; ADEC 2001b). A summary of the inventory for 1995 and 2001 is shown in Table 2-7. This inventory is for a typical winter day, because exceedances are more likely to occur then. ADEC (2001a) also provides emissions reduction estimates for various control strategies that the borough has implemented to come into attainment with the NAAQS for CO. A summary of estimated emissions reductions is shown in Table 2-8.

The onroad mobile source emissions inventory was developed with a hybrid approach. Because of the importance of cold-start emissions in the borough, onroad mobile source emissions are separated into initial-idle emissions (when the vehicle starts up and idles before traveling) and traveling emissions. Estimates of initial-idling emissions were based on emissions testing of vehicles during winter in Fairbanks. Later onroad emissions (traveling emissions) estimates were based on EPA's Serious Area CO Model (Darlington and Kahlbaum 1998), which is a version of EPA's Mobile Source Emissions Factor (MOBILE) model that accounts for cold-temperature emissions standards.[10] The remainder of the CO emissions inventory (nonroad, area, and point sources) was developed with the emissions factors described

---

[10] A new version of the MOBILE model, MOBILE6, has been recently released for general use. Cold-temperature emissions standards are incorporated into this model. Thus, MOBILE6 will eventually replace the Serious Area CO Model.

in EPA (1998a) and activity factors estimated locally, except for aircraft and related emissions, which were estimated with the Federal Aviation Administration Emissions and Dispersion Modeling System.

As shown in Table 2-7, onroad motor vehicles are estimated to have contributed about 69% (21.7 tons/d, or tpd) of total CO emissions in 1995. By 2001, these emissions are estimated to have declined by 7.3 tpd, to 14.4 tpd (62% of total CO emissions), even though VMT in Fairbanks continued to grow. The fraction of CO emissions attributed to onroad motor vehicles is higher in Fairbanks than in the national inventory (see Table 1-2) reflecting the large contribution of cold-start emissions in Fairbanks. Although cold-start, initial-idle, and traveling emissions are estimated to have decreased from 1995 to 2001, the fraction of onroad mobile-source emissions attributed to cold-start and initial-idle emissions has increased. In 1995, cold-start and initial-idle emissions were estimated to be 38% of total onroad emissions; by 2001 the estimate had increased to 45%.

Other emissions sources in Fairbanks include nonroad, area, and point sources. In 2001, nonroad sources are estimated to contribute about 16% (3.7 tpd) of total CO emissions. Area and point sources are estimated to contribute about 4% (0.9 tpd) and about 19% (4.3 tpd), respectively. Although other emissions sources (onroad, nonroad, and area sources) have generally decreased from 1995, point sources have increased by 5%. The contribution to enhanced surface CO concentrations from large point sources, particularly those that release emissions from tall stacks, is not well known. Surface inversion conditions can inhibit mixing of those emissions down to the surface, reducing the impact of nearby point sources. However, their contribution cannot be completely discounted because the surface inversion can break down, enabling rapid mixing with the surface, or the trapped pollution can be transported to the surrounding uplands.

Overall CO emissions in the borough were estimated to have declined by about 8.1 tpd from 1995 to 2001. As shown in Table 2-8, 90% of the reductions (7.3 tpd) came from onroad motor vehicles. The largest contributor to the reduction was fleet turnover, the replacing of older vehicles with new ones that have more stringent emissions standards. The next largest contributors to the reductions are the inspection and maintenance (I/M) program (both operation and enforcement) and expansion of the number of parking spaces equipped with electric outlets for plug-in units. Area-source emissions reductions also occurred as a result of reduced wood-burning. Reductions in nonroad mobile-source emissions are attributed to reduced aircraft emissions.

TABLE 2-7 CO Emissions Inventory for the Fairbanks North Star Borough Nonattainment Area

Source Category	1995 Emissions (tons per day)	2001 Emissions (tons per day)
Onroad sources	21.69	14.40
Cold-start and initial-idle emissions	8.28	6.49
Traveling emissions	13.41	7.91
Nonroad sources	4.00	3.66
Airport ground support equipment	2.36	1.91
Aircraft, excluding ground support equipment	1.27	1.37
Snowmobiles	0.27	0.28
Railroad operations (Locomotives)	0.04	0.04
Forklifts	0.02	0.03
Air compressors	0.01	0.01
Area sources	1.53	0.89
Residential wood burning	1.29	0.67
Fuel oil	0.16	0.13
Natural gas	-	0.01
Structural fires	0.08	0.08
Point sources	4.14	4.33
MAPCO (Williams)	0.15	0.43
Fort Wainwright	2.09	1.72
GVEA/North Pole	0.02	0.02
University of Alaska (Fairbanks)	0.51	0.52
Petro (Star)	0.01	0.01
Fairbanks MUS (Aurora)	1.37	1.63
Total	31.36	23.29

Source: ADEC 2001a.

Increased operations at the local electric utility and the refinery are the reason for increased point-source emissions. Using the statistical rollback approach described in Chapter 1, ADEC (2001a) concluded that the overall 26% decrease in emissions will be sufficient for attainment.

TABLE 2-8 Summary of Emissions Reductions in the Borough Over the 1995-to-2001 Time Period

Source Category	Reduction (tons per day)	% Change
Onroad Mobile Sources		
Improved New Vehicle Emissions Standards (fleet turnover)	5.73	
I/M Program Enhancements	1.26	
I/M Increased Enforcement	0.25	
Expanded Plug-Ins	0.05	
Total Onroad Mobile Sources	7.29	33.6%
Area Sources	0.64	41.8%
Nonroad Mobile Sources	0.34	8.5%
Point Sources	-0.19	-4.6%
Total Reductions	8.11	26%

Source: ADEC 2001a.

Vehicle-Fleet Characteristics

Characterizing the onroad fleet during design conditions is critical for estimating motor-vehicle emissions. Figure 2-9 shows CO emissions rates predicted by EPA's MOBILE5b model for vehicle emissions in 2000 for the last 25 model years of automobiles and light-duty trucks using the default vehicle-fleet distribution. The higher emissions rates for earlier-model-year vehicles result from less stringent certification standards when the older vehicles entered the fleet and deterioration of their emissions-control systems over time. However, older vehicles tend to be relegated to being secondary vehicles and are driven less because they are generally less reliable than the other vehicles in the household. The greatest contribution to overall emissions in most urban areas tends to come from the middle-aged vehicles, which have high emissions rates, are in the fleet in large numbers, and are still driven extensively.

Local fleet composition affects emissions from the onroad fleet. The Fairbanks onroad-fleet composition during the winter months depends less on vehicle ownership than on vehicle use. About 8% of vehicles registered in the borough obtain seasonal waivers; that is, they are not operated between November 1 and March 31 and are exempted from the I/M program. Vehicles

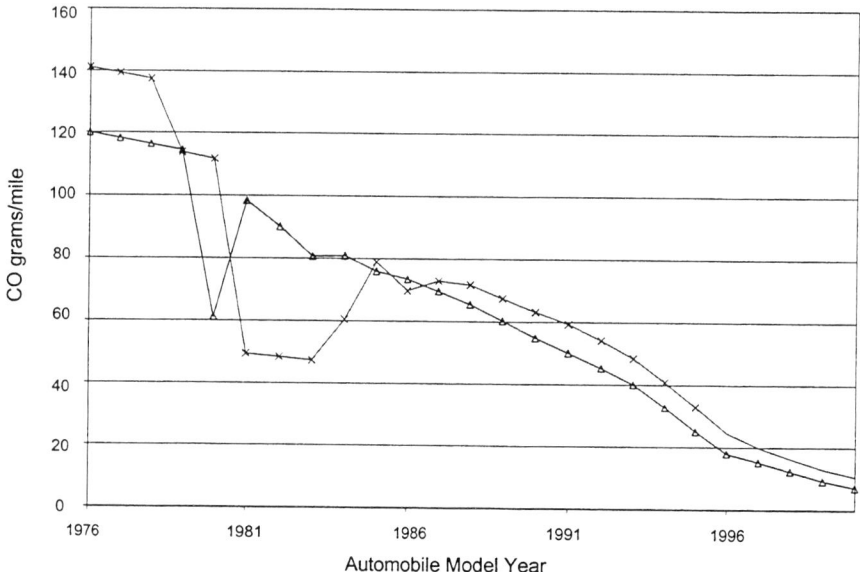

FIGURE 2-9 Predicted MOBILE5b CO basic emissions rates in 2000 (at 20°F, average vehicle speed of 15 mph, national default fleet assumptions, and no I/M program). The counterintuitive dips in emissions rates, such as those for 1980-model-year automobiles, reflect internal model correction factors in MOBILE5b that apply differently to certain model-year technology groups.

operated in the borough during extremely low temperatures are typically newer and better-maintained than the average summer fleet.

In the winter of 2000, ADEC conducted a license-plate survey in downtown parking lots to ensure that only the vehicles normally used during the winter were included in emissions modeling for state implementation plan (SIP) development (ADEC 2001b). Large differences in model-year distributions can lead to much larger changes in emissions because of the nonlinear relationships involved. Given the importance of using accurate vehicle fleet compositions in modeling, ADEC should undertake a new round of winter fleet studies. The agency is encouraged to address the following in the experimental design to ensure that representative samples are obtained: land use at license-plate sampling locations (which is related to trip types), potential interactions between temperature and fleet composition, collection of supplemental onroad license-plate data, and potential inaccuracies in the vehicle-registration database.

Fairbanks Case Study 77

CO CONTROL PROGRAMS

This section examines the motor-vehicle CO emissions-control programs that have been or may be implemented in the borough. In general, mobile-source emissions-reduction strategies can be categorized into new-vehicle certification programs, fleet-turnover incentives, in-use vehicle control strategies, and transportation control measures (TCMs) (Guensler 1998, 2000[11]). Except for new-vehicle certification standards, which are discussed in Chapter 1, this section outlines various emissions-control strategies in each major control category that are applicable to the borough.

Fleet-Turnover Incentives

Strategies designed to increase vehicle-fleet turnover will increase the number of clean vehicles entering the fleet each year while retiring older, high-emitting vehicles. These include voluntary scrapping programs, in which an agency purchases and scraps high-emitting vehicles, and economic incentives, in which taxes and other fees are designed to encourage the ownership of newer vehicles over older ones. Fairbanks is already one step ahead of most urban areas because its flat registration fee applies to all vehicles. This policy could be enhanced by eliminating registration fees for the first few years of new-vehicle ownership or by adding a surcharge to fees for older, high-emitting vehicles.

In-Use Vehicle Controls

Clean-fuels programs and I/M programs are the mainstays of nationwide in-use vehicle controls. In Fairbanks, where cold engine starts contribute substantially to the regional nonattainment emissions inventory, cold-start controls in the form of engine preheating also provide valuable emissions reductions.

---

[11]Guensler, R. 2000. TRANS/AQ 2000; Transportation and Air Quality; Courseware CD-ROM; Version 1.0; Georgia Institute of Technology; Atlanta, GA; August. http://transaq.ce.gatech.edu/

Fuels Control

Three major fuels-control programs have the potential to reduce CO emissions in the borough: increased fuel volatility, use of oxygenated fuels, and reduced fuel sulfur concentrations. Detergent additives also help to control emissions by keeping fuel injectors clean and keeping emissions-control systems functional.

Gasoline Volatility

Gasoline volatility is critical to cold starting. Higher fuel vapor pressures ensure adequate fuel vaporization during cold starting. However, when vapor pressures are too high, vapor lock can prevent liquid fuel from flowing to the fuel injectors. Volatility standards are designed to ensure that engines can start in cold weather while minimizing both the possibility of vapor lock and evaporative emissions from the fuel tank under normal operating conditions. During the winter months (middle of September to middle of May), refiners can produce gasoline with a maximal vapor pressure of 15 psi. Both refineries that supply Fairbanks report that their gasoline in winter has a vapor pressure of 14.5 psi (Boycott and Cherry 2001; Henderson 2001).

Oxygenated Fuels

Adding an oxygenate to the fuel increases the oxygen-to-fuel ratio in the combustion process, changing the combustion chemistry and decreasing emissions of CO formed during incomplete combustion. Colorado instituted the use of oxygenated fuels (oxyfuels) in 1988 to reduce winter CO concentrations. EPA later extended the oxyfuels program to other areas of the United States that were exceeding the NAAQS for CO (typically during winter). The federal oxyfuels program required that fuels contain an oxygenate—usually methyl *tertiary*-butyl ether (MTBE) or ethanol—with oxygen at 2.7% or more by weight. A 1997 study of the winter oxyfuels program initiated by the White House Office of Science and Technology Policy concluded that at temperatures above 50°F, CO emissions from most vehicles were reduced by about 3-6% per weight percent oxygen (NSTC 1997). Emissions reductions were generally smaller in newer-technology vehicles (those with closed-loop fuel control and three-way catalysts) and larger in high-emitting and older-technology vehicles (those with permanent open-loop fuel control and two-way catalysts).

Oxyfuels with MTBE were introduced in Fairbanks in October 1992. Public complaints about the odor and potential health concerns led to a suspension of and then statewide ban on the use of MTBE in Alaska in December 1992. Fairbanks has been reluctant to reintroduce oxyfuels, such as gasoline containing 10% ethanol, despite the fact that an ethanol blend is required in Anchorage and provides the largest source of projected CO reductions there (ADEC 2001c).

Because extremely high emitters are less likely to operate in Fairbanks during winter (because of reduced reliability) and oxygenates provide little or no benefit for late-model vehicles under warmed-up running conditions, the reintroduction of oxyfuels may not provide a large emissions-reduction benefit for running exhaust emissions in Fairbanks. However, the use of oxyfuels may reduce CO emissions during cold starts in late-model vehicles when the O_2-sensor is off line during the first several minutes of operation. There is a lack of information on the effectiveness of oxyfuels at temperatures below 50°F. EPA (Mulawa et al. 1997) tested three vehicles at 20, 0, and -20°F at its cold-weather facility using unleaded gasoline containing 10% ethanol (3.5% oxygen by weight). Two cars showed substantial improvement in CO emissions, and a third showed no emissions benefits. The Colorado Department of Public Health and Environment (Ragazzi and Nelson 1999) found an average 11% decrease in CO emissions from switching to 10% ethanol blended fuels in 24 vehicles that it tested at 35°F. Although theory suggests that oxygenated fuels should provide emissions benefits under extreme cold-start conditions, available data are not sufficient to support or refute the argument. Before an oxygenate, such as ethanol, is required in Fairbanks gasoline during the critical mid-winter months, more research on its effectiveness at low temperatures should be conducted. However, such studies are warranted and should be conducted in the near term.

BOX 2-3 Recommendations: Oxygenated Fuels

Alaska, EPA, and others should conduct additional research and vehicle testing to assess the effectiveness of ethanol in gasoline for decreasing CO and air-toxics emissions during cold starts and operation at ambient temperatures below 20°F. If such research indicates that substantial benefits can be achieved, ethanol blending should be considered for Fairbanks.

Sulfur Content

A key finding of the Auto-Oil Project (Benson et al. 1991) was that fuel sulfur increases exhaust emissions. Sulfur in gasoline is known to affect the efficiency of vehicle emissions-control systems adversely by poisoning the catalyst, thus decreasing pollutant conversion efficiency and potentially lengthening the time needed after vehicle ignition for the catalyst to become effective. The 1991 Auto-Oil study concluded that reducing sulfur concentrations from 450 to 50 ppm would result in a 13% decrease in CO exhaust emissions in 1990 Tier 0 technology vehicles. In addition, low-sulfur fuel is expected to reduce emissions of hydrocarbons (HC), nitrogen oxides (NO_x), hydrogen sulfide, sulfur dioxide, and sulfuric acid aerosols, as well as lengthen lubricant and engine life. Reversing the effects of sulfur on catalytic performance requires fuel-rich conditions and aggressive accelerations that achieve high catalyst temperatures (about 1,200°F). However, sulfur's effects are not easily reversed on the newer-model lower-emissions vehicles (Truex 1999). To guard against the poisoning effects of sulfur, it is best to operate these newer-model vehicles on low-sulfur fuel only. On the basis of concerns about the increased sensitivity of the newer-technology vehicles to sulfur poisoning, EPA published new fuel standards requiring refiners to meet an average sulfur concentration of 30 ppm beginning January 1, 2006 (EPA 2000c). Additional studies are needed on the effect of high-sulfur gasoline on catalyst efficiency and light-off time (the time it takes the catalyst to reach peak efficiency after starts) in cold climates.

Gasoline available in Fairbanks comes from the Tesoro and Williams refineries. Tesoro's average fuel sulfur concentration is less than 1 ppm (although a particular batch may have up to 5 or 10 ppm because of batch-to-batch variations), and Williams's average fuel sulfur is around 200 ppm. Immediately requiring the use of low-sulfur gasoline during winter months in Fairbanks is likely to reduce CO emissions. Reduction of gasoline sulfur on a year-round basis would be even more effective. However, adequate safeguards are needed to ensure that gasoline prices in Fairbanks do not increase substantially and that other areas of the state are not negatively effected. Introduction of lower-sulfur gasoline could be facilitated through accelerated approval of refinery-construction permits or through a state-brokered gasoline-exchange program. Policy and economic analyses should be conducted in consultation with the two local refiners to determine the best approach.

> **BOX 2-4** Recommendations: Fuel Sulfur Content
>
> The borough should consider requiring the sale of low-sulfur gasoline as soon as possible. Introduction of low-sulfur gasoline could be facilitated through accelerated approval of refinery construction permits and through a state-brokered gasoline-exchange program. Policy and economic analyses, in consultation with the two local refiners, are needed to determine the best approach to ensure that this mandate will not substantially increase the cost of gasoline to Fairbanks residents or compromise the air quality in other parts of the state. A public-awareness campaign to explain the benefits of low-sulfur fuels is needed, and the sulfur content of fuels should be posted at gasoline stations.

I/M Programs

Inspection and maintenance (I/M) programs have been instituted in many jurisdictions to ensure that emissions-control equipment operates efficiently throughout the life of a vehicle. The programs attempt to identify vehicles that have higher emissions than allowable and to ensure that such vehicles are repaired or removed from the fleet. Inspection typically involves regularly scheduled exhaust tests that measure emissions of CO, HC, and sometimes NO_x. I/M tests may also include visual inspection of the components that control evaporative and exhaust emissions and a functional gas-cap test.

The borough implemented an I/M program in 1985 to reduce motor-vehicle CO emissions. The program includes exhaust, visual, and functional tests of model-year 1975 and newer vehicles. The borough operates a test-and-repair program in which tests are performed at service stations that can also do repairs. Exhaust emissions are evaluated with a two-speed idle test. Vehicles less than 2 y old and vehicles that are not driven during the winter (from November 1 to March 31) are exempted from the I/M test. Vehicle registration along with windshield stickers are used to distinguish between vehicles driven in winter, which are subjected to the I/M test, and vehicles that are not. Until 1997, annual I/M tests were required; in 1997, Alaska passed a law changing testing to every 2 y.

As of July 1, 2001, model-year 1996 and newer vehicles must pass a test of their onboard diagnostics (OBDII) system in addition to the standard tailpipe test. If the test is operating properly, OBDII inspections will fail vehicles if the emissions-control components are or have been malfunctioning or if the sensors monitoring emissions-control components are malfunctioning. In Fairbanks, OBDII testing programs can aid in identifying problems that occur under a variety of operating conditions, not only when a vehicle is being idle-tested. NRC (2001) contains a discussion of major issues associated with OBDII I/M programs.

The I/M program is a critical element of the borough's efforts to control CO. The SIP submitted by Alaska specifies that the largest reductions in CO emissions from local initiatives in Fairbanks will be from improved vehicle testing and increased enforcement of the I/M program. The borough estimates that the I/M program is effective in reducing vehicle CO emissions; the program is expected to reduce fleet-average CO emissions rates by 15.1% compared with the modeled emissions without an I/M program. That reduction is almost three times the 5.4% reduction estimated for the federal I/M performance standard for a similar program (ADEC 2001a). However, the move from annual to biennial testing probably has decreased the overall effectiveness of the program; failure rates went from 10% in the annual program to 14% in the biennial program (Lyons 2001). The committee found no evidence that the benefits of the I/M programs had been evaluated with respect to in-use vehicle emissions.

Methods and Data for Evaluating the I/M Program

The borough would benefit from continuing evaluation of its I/M program. The evaluation should include study of the overall emissions benefits of the I/M program using in-use vehicle emissions data, the level of program noncompliance, the types and effectiveness of emissions-related vehicle repairs, and the adequacy of current types of testing for identifying high-emitting vehicles. Various data could be used to estimate vehicle emissions, repairs, and noncompliance, including I/M-program data, remote-sensing data, and data collected through special emissions studies. EPA has described several methods for using this information to estimate effects of I/M programs in emissions (Sierra Research 1997; EPA 1998b, 2001d, 2001e). Many of the issues associated with I/M evaluations are discussed in a recent NRC report (NRC 2001) and in evaluations completed by state and independent researchers (CARB 2000; IMRC 2000; Stedman et al. 1997; Wenzel 1999). The NRC report concluded that each method and data source for evaluating I/M programs had its own inherent advantages and disadvantages. The report further concluded that a thorough understanding of all potential effects of I/M on vehicle emissions would come only through the use of multiple data sources and evaluation methods.

The committee believes that ongoing evaluation of the I/M program will help identify ways to improve it. For example, Sierra Research (1999) indicated that 1984 and older gasoline-powered cars and trucks accounted for only 8.2% of the total vehicle-miles traveled but for 20.7% of the CO emitted from the light-duty gasoline fleet. An evaluation of its program could help the borough to determine whether modified or additional I/M programs should be adopted to ensure compliance and reduce emissions from older vehicles. The evaluation also might indicate that mandatory replacement of O_2 sensors in older fuel-injected vehicles or inspection of pre-1975-model-year vehicles could provide significant benefits.

Remote sensing is one possible method of collecting emissions data to evaluate the I/M program. It is used to measure emissions from individual vehicles as they drive by a roadside sensor. Other advances that have facilitated the collection and interpretation of remote-sensing measurements include pattern-recognition software to read vehicle license plates automatically and sensors to measure speeds and accelerations of passing vehicles. Because extreme cold can affect system performance, remote-sensing systems are typically not deployed in the winter. However, remote sensing is currently deployed year-round as a method for screening low-emitting vehicles from scheduled I/M tests in Greeley, Colorado (Klausmeier and McClintock 1998) and has been deployed in Yellowstone National Park (Bishop et al. 1999) in the winter to measure emissions from snowmobiles. An alternative approach may be to deploy the remote-sensing system solely to read vehicle license plates automatically and produce an improved breakdown of vehicle types and model years being operated in high CO areas.

Remote-sensing data can be used to evaluate an I/M program by using the license-plate data collected at the same time as emissions are being measured, and existing registration and I/M records. With this information, how quickly repair effectiveness diminishes, how much repair takes place before the I/M test, and the number and emissions of vehicles that avoid testing can be estimated. EPA (2001d) has issued draft guidance on the use of remote sensing for evaluating I/M programs. However, implementing remote sensing for I/M-program evaluation or testing requires careful attention to issues associated with site selection, quality control, and vehicle operating conditions and engine load. Those issues are discussed in NRC (2001).

BOX 2-5 Recommendations: I/M Programs in the Fairbanks North Star Borough

Frequency and Exemptions

The borough should consider resuming annual inspections. The committee is aware that this may require state legislation. The borough should also expand the coverage of its I/M program to include 1968-1974 model-year vehicles. The current new-car testing exemption is reasonable; it may also be cost-effective, starting with the 2000 model year, to expand the exemption to cover the four most recent model years.

Improvements in Emissions Testing

The borough should comprehensively assess emissions-testing methods to determine appropriate inspection procedures for various vehicle technologies. This assessment should consider the use of annual two-speed idle tests for pre-1982 vehicles and biennial testing under driving load conditions for 1982-1996 vehicles. The assessment should also consider the issues associated with using OBD testing in cold climates. Because of the frequency of O_2-sensor failure, the borough should also evaluate the potential emissions-reduction effectiveness of a mandatory O_2-sensor replacement program for older, high-mileage vehicles and implement such a program if it is found to be effective.

Remote Sensing

Use of remote-sensing capabilities should be considered for the borough's I/M program, as temperatures and atmospheric conditions permit, to help to characterize emissions of the vehicle fleet. A continuing remote-sensing program should be considered to evaluate the potential effectiveness of the I/M program, as is done in other regions. The borough could also consider using remote sensing to identify vehicles that must be tested or other vehicles that could be given an exemption.

(Continued)

> **BOX 2-5** *Continued*
>
> *Ongoing Evaluation of I/M*
>
> The borough should evaluate the I/M program more rigorously to estimate its emissions-reduction benefits and to identify where improvements to the program are needed. The evaluation should allow the direct comparison of the emissions reductions achieved by the program with those estimated in the borough's attainment plan. It should also look for methods to improve the effectiveness of I/M.

Cold-Start Controls

Emissions tests conducted in Fairbanks indicate that cold-start and onroad CO emissions can increase by an order of magnitude when ambient temperature goes from 75°F to -20°F (Sierra Research 1996). The increase reflects the effects of ambient temperature on the duration of fuel-rich conditions during cold starting before closed-loop fuel metering occurs and a substantial delay in reaching catalyst light-off temperature. As described earlier, a large component of vehicle emissions during winter is attributed to cold-start operations. Therefore, effective local in-use vehicle control strategies can focus on reducing the number of winter engine cold starts or the length of time it takes for vehicles to warm up. The borough has provided free bus service during the winter to reduce the number of cold starts and has focused extensive efforts on plug-in cold-start controls to reduce the time it takes vehicles to warm up.

Plug-In Strategies

Starting and operating a car or truck in very cold climates has long been a challenge. Electric heating of the engine before starting is a part of everyday life during the winter in Fairbanks. Without preheating, some vehicles will not start under the most extreme cold conditions. Vehicles that operate in the winter in Fairbanks need to have one or more electric heating devices (plug-ins) installed. Residents plug the devices into electric outlets at home, at work, and in other places. The most popular device is a 600-watt electric resistance heater submerged in the engine coolant. On larger engines, a second coolant heater may be used. Electric heating pads are also glued to the underside of the oil pan to help keep the lubricant warm. The battery is sometimes wrapped in an electric heating blanket to improve cold-cranking performance. Winterizing a vehicle with two 600-watt block heaters and an oil-pan heating pad is estimated to cost several hundred dollars, including parts and labor. Indeed, winterizing is considered part of the cost associated with owning a car in a cold climate.

Engine preheating substantially reduces CO and HC emissions during cold start and engine warmup. In a study that estimated cumulative CO emissions from six vehicles in a cold start plus 15 min of idling, the ADEC (2001b) reports a 70% reduction in CO emissions after plugging in the vehicles for 4-8 h. Vehicle preheating provides side benefits, such as lower engine wear, quicker passenger-compartment heating, and windshield defrosting. For the temperatures when CO violations are most likely (0-20°F), preheating is not required to start the engine, and most plug-in and extended idling activity is probably undertaken for personal comfort. Therefore, the borough has encouraged people to plug in even when the temperature is as high as 20°F.

The challenges to plug-in programs are twofold: making sure that active electric outlets are available wherever vehicles are parked for any length of time and getting people to use the plug-ins even when it is warm enough (0-20°F) for vehicles to start without preheating. Many businesses, schools, and government offices provide electric plug-in outlets in their parking lots, but others do not. The borough actively requires employers and government buildings to provide more parking spaces with plug-in receptacles and the necessary electric power. A recent ordinance implemented by the borough requires employers and businesses with more than 274 parking spaces with outlets to provide power when the ambient temperature is less than 21°F (Ordinance No. 2001-17). Although seen as a convenience by motorists to reduce the need for extended idling of unoccupied parked cars, the program is intended to reduce engine-start emissions. Additional studies on individual responses in terms of the frequency and duration of plug-in use would help the borough to determine the effectiveness of the current strategies and to determine whether plugging in should be mandatory and integrated into parking regulations with appropriate enforcement.

Many uncertainties are associated with the use of plug-ins to reduce cold-start emissions. Relationships among engine size, heater power, and heating time required to reduce cold-start emissions substantially are not well understood. The time for engine idle between cold starting and driving off may also be an important factor in total CO emissions during a vehicle trip. Moderate idling times will increase idle emissions but may reduce drive-off emissions, resulting in a net decrease in emissions for the entire trip. Sparse data suggest that 5 min of idle after an engine cold start under very cold conditions may be near the optimum for reducing total CO emissions for a trip, but this optimum time varies among vehicles depending on the emissions-control system (Sierra Research 1999). In addition, some vehicle models equipped with an air pump may automatically divert air from the catalyst to protect the catalytic material from overheating and damage during extended engine idling. That can increase idle rates of CO emissions substantially (Sierra Research 2000). Additional study with instrumented vehicles at low temperatures (-20°F to 20°F) would help to determine the potential effects of extended vehicle idling on emissions.

After-Market Retrofits

Retrofits to existing emissions-control systems to reduce cold-start CO emissions are also possible. Catalytic converters can be preheated to reduce light-off time, thereby reducing cold-start emissions. Electrically heating the catalyst just before engine starting has shown promising reductions in HC and CO. The greatest effect was realized when supplemental air was also injected into the catalyst inlet during open-loop operation (Heimrich 1990; Heimrich et al. 1991); however, no studies have been conducted on how this strategy might be practically implemented as an after-market control measure. Advances in that technology have reduced the substantial electric power demand for catalyst heaters, but it remains doubtful that in-fleet vehicle batteries can both heat the catalyst and then start the engine at low temperatures. Researchers could explore the value of a separate plug-in system for the catalyst. Another potential retrofit would be the addition of air into the exhaust manifold (upstream of the catalytic converter) during fuel-rich open-loop operation, although the introduction of air at temperatures below about 0°F may lengthen the time needed for full catalyst warmup. The required modifications to the exhaust, electric, and other emissions-control components needed for retrofit options may be extensive and more suitable for the original equipment manufacturer.

> **BOX 2-6** Recommendations: Vehicle Plug-ins
>
> The borough should continue to expand the plug-in program by requiring or encouraging the equipping of more parking spaces with electric outlets for plug-ins. Efforts to increase the use of plug-ins at 0-20°F are especially warranted. Public-education campaigns should continue. Adoption and enforcement of engine-preheating regulations on days expected to have high ambient CO concentrations should be considered. However, further analyses could determine the factors that motivate the voluntary use of plug-ins and the incentives that will expand their use. Additional effort should be directed toward understanding the relationships among engine size, heater power, and the heating time required to substantially reduce cold-start emissions.

Transportation Control Measures

Transportation control measures (TCMs) are actions designed to change travel demand or vehicle operating characteristics to reduce motor-vehicle emissions, energy consumption, and traffic congestion. TCMs include transportation-demand management (TDM) strategies and transportation-supply improvement (TSI) strategies. TDM strategies attempt to reduce the frequency or length or shift the timing of automobile trips by changing driver behavior via regulatory mandates, economic incentives, and education campaigns. In contrast, TSI strategies attempt to reduce emissions by changing the physical infrastructure of the road system to improve traffic flow and reduce stop and go movements. Table 2-9 describes TCMs that have been applied in urban areas nationwide (EPA 2001f) and notes whether they have been implemented in Fairbanks.

The Alaska Department of Transportation and Public Facilities, in conjunction with the borough, has implemented a few TCMs, including a package of highway and intersection improvements to reduce traffic delays, a public-education campaign to encourage the use of plug-ins, and a program to promote mass-transit use by making buses free during the winter. Only small CO emissions reductions are attributed to the programs in the SIP (ADEC 2001a). Additional improvements in the road system, a coordinated mass-transit program, and the purchase of new buses are listed as contingency

TABLE 2-9 Transportation Control Measures (TCMs)

TCM Description	Applicability to Fairbanks
Improved public transit	
Incentives for single occupancy vehicle commuters to use convenient and reasonably priced mass transit alternatives. The three major ways of increasing ridership on public transit are (1) system/service expansion, (2) system/service operational improvements, and (3) inducements to increase ridership.	· Public bus system since 1977. · Free bus rides during winter CO season (11/1 to 3/31) started in 2000. Extensive campaign informing residents about free rides and encouraging them to try transit. · Coordinated transit program and new buses considered in SIP.
Traffic flow improvements	
Strategies that enhance the efficiency of a roadway system, without adding capacity, including traffic signalization, traffic operations, and enforcement and management.	· Highway and intersection improvements to reduce traffic delays implemented since 1996. · Other roadway improvements considered in SIP.
High-occupancy vehicle (HOV) lanes	
Roadways dedicated for HOV use.	· Not adopted.
Intelligent transportation systems	
Traffic detection and monitoring, communications, and control systems. Examples include traffic signal control, freeway and transit management, and electronic toll collection systems.	· Not adopted.
Bicycle and pedestrian programs	
Includes sidewalks, bicycle lanes, and bicycle racks.	· Available along major roads but not used much during CO season.
Commute alternative incentives	
Incentives, usually employer based, to encourage commuters to carpool or use transit services.	· Not adopted.
Telecommuting	
Working at home using electronic communication instead of physically traveling to a distant work site.	· Not adopted. · Could be considered for alert days.
Guaranteed ride home programs	
Ensures transportation (e.g. taxi or transit passes) for carpooling employees in the case of an unforeseen circumstance.	· Not adopted.

(Continued)

TABLE 2-9 Continued

TCM Description	Applicability to Fairbanks
Work schedule changes	
Adjusting hours worked to control peak emissions: Examples include staggered hours, flextime, and a compressed work week.	· Not adopted. · Could be considered for alert days.
Trip reduction ordinances (Regulatory mandates)	
Regulations that attempt to adjust personal travel decisions through employer-based incentive/disincentive programs.	· Not adopted. · Could be considered for very large employers.
Congestion pricing	
Financial disincentives to driving on highly-used roadways, or priced alternatives to a congested roadway.	· Not adopted.
Parking pricing	
Programs that encourage single-occupant vehicle users to switch to other means of travel by imposing fees for parking, or that encourage shifting times for vehicle starts away from peak CO periods.	· Not adopted. · Could be considered to shift cold starts away from peak CO times.
Parking management	
Allocation of parking spaces intended to encourage single-occupant vehicle users to use other means of travel.	· Not adopted.

Source: EPA 2001f.

measures if the borough exceeds the CO standard (ADEC 2001a). It is unlikely, however, that the borough could achieve significant additional reductions in CO emissions through those measures. Most of the other TCMs recommended by EPA (Table 2-9) are probably not appropriate for Fairbanks, given the low levels of congestion, low density of development, and severe winter weather conditions. Some of the demand-management and supply-improvement strategies that may be applicable to Fairbanks are discussed below.

Transportation-Demand Management

Demand-management measures include no-drive days, employer-based trip-reduction programs, parking management, park-and-ride programs, work-schedule changes, mass-transit fare subsidies, and public-awareness programs. Such strategies can be categorized into trip-reduction mandates, market incentives, voluntary programs, and education-exhortation campaigns (Guensler 1989). In general, trip-reduction mandates have not proved effective in the United States at reducing vehicle-miles traveled (Guensler 1998). However, such mandates may be effective for very large employers, such as universities and major government centers.

Market incentives, or strategies that attempt to influence travel decisions by adjusting the prices of travel modes, have been applied in Fairbanks to encourage bus ridership. The borough established a public-transit system, the Metropolitan Area Commuter System, in 1977. According to the borough Transportation Department, ridership has varied from 550,000 per year during the 1980s to fewer than 200,000 riders per year in 1988, when several important routes were eliminated because of budgetary constraints. Ridership increased to about 230,000 riders per year after two new routes were added in 1996. In 2000, the borough received federal Congestion Mitigation and Air Quality funds to begin implementing a free-ride program during the winter season (November 1-March 31). This program included an extensive public-information campaign to inform residents about the free rides and to encourage them to try public transit. Ridership increased by 72% while the free-ride program was in effect, though this was estimated by ADEC (2001a) to provide a rather small 0.05-tpd reduction in CO emissions. Aside from these numbers on past ridership, the borough has not conducted studies of transit needs in Fairbanks. Such a study would help identify whether expanding the service could result in significant emissions reductions.

Recent studies indicate that one of the most cost-effective TCMs is associated with parking-price programs (see, for example, Guensler [1998] for additional details). A possible application in Fairbanks is the implementation of graduated parking pricing in a downtown parking garage that is under construction. Under such a program, the price for parking could be used to provide incentives for reducing or eliminating vehicle starts during late afternoon hours (when observed CO concentrations are highest). For example, parking would be less expensive (or free) for those not leaving the lot between 4 and 7 p.m. Research on effects on travel behavior would need to be performed to make sure that not everyone tries to leave at 3:59 p.m. or chooses to use other parking facilities. This strategy would clearly have a greater effect if expanded to other parking facilities and made part of an alert-day strategy.

Another set of TCMs that needs further exploration in Fairbanks includes telecommuting and teleservices. Many jobs are conducive to telecommuting (for example, working from home rather than commuting to the office) at least

1 d per week. Many state and local governments have worked with employers to establish telecommuting programs for their employees. Teleservices, the provision of government or business services over the phone or the Internet, might also be encouraged and expanded. Recent statistics on computer and Internet access in Alaska support the feasibility of such programs. In 2000, almost 65% of households in Alaska had computers, and over 55% had Internet access (U.S. Department of Commerce 2000). Participants in such programs realize many benefits, including reductions in gasoline use, in vehicle wear, in time, and in hassles of cold-weather driving.

Some TDM measures might provide short-term emissions reductions for Fairbanks if implemented through an expanded alert-day program when conditions are likely to create CO exceedances. Although CO alert-day programs would not have a large effect on average annual CO emissions, they might have enough effect to help to avert an exceedance. For example, on alert days, employees could be encouraged to leave work earlier or later, shifting vehicle emissions away from the period when maximum concentrations are most likely. On the same days, the borough might encourage residents to put off trips in the nonattainment area for services or shopping. The EPA's Transportation Air Quality Center provides a database of previously implemented programs of this sort and ways to evaluate their effectiveness (EPA 2002).

The success of alert-day programs depends on two factors. First, the borough must be able to identify conditions that are conducive to CO exceedances with a reasonable degree of accuracy. As discussed earlier and shown in Table 2-4, data from the borough show that over a 3-y period staff predicted four exceedances that occurred, did not predict three that occurred, and predicted 12 that did not occur. Exceedances, however, need not be perfectly predicted. The absence of an exceedance on an alert day could be due to the emissions-reduction efforts on the part of the community rather than to a forecasting failure. Second, local employers need to cooperate in the alert-day programs, allowing their workers to stay late or telecommute, for example. If voluntary participation falls short of the required levels, the borough could consider adopting modified trip-reduction or mandatory plug-in ordinances.

Transportation-Supply Improvement

Transportation-supply improvement (TSI) strategies attempt to reduce emissions by improving traffic flow, usually by increasing the effective capacity of the existing roadway system. Because CO emissions from vehicles are higher under hard acceleration conditions, improvements in traffic flow can help to reduce them. Given the low levels of congestion and large excess transportation capacity in Fairbanks, most TSI strategies implemented in other urban areas would not help to reduce CO emissions in Fairbanks. Since 1996, Fairbanks transportation planners have completed 11 highway-improvement projects in the nonattainment area, including several intersection improvements. ADEC estimates that these projects increased average speeds in the nonattainment area by about 0.2 mph. Although the modeling results show only a slight increase in average speeds, intersection projects have the potential to reduce CO emissions substantially by improving traffic flow and reducing the idling and acceleration associated with intersection delays (Hallmark et al. 2001). Signal timing improvement should be a priority for implementation in downtown Fairbanks and should optimize emissions reductions rather than focusing solely on total delay.

BOX 2-7 Recommendations: Traffic Flow and Motorist-Directed Control Strategies

The borough should explore parking pricing, telecommuting, and tele-services strategies. The borough should evaluate the effectiveness of its "alert-day" program and consider enhancing it. In addition, a travel-demand study, including a winter diary and a transit-ridership survey, should be undertaken to provide a basis for evaluating the potential effectiveness of proposed TCMs.

Public Education and Surveys

Education and exhortation programs can increase the success of control strategies, especially voluntary ones, which depend on consumer behavior. The borough has adopted some control strategies in which consumer participation is voluntary. Public-education efforts were initially aimed at increasing awareness about the CO problem and how the use of plug-ins and mass-transit may help to alleviate the problem. Paid television and radio announcements were aired during heavy viewing or listening times. Later messages explained the increased enforcement activities associated with the I/M program and the possible adverse consequences for the borough of not reaching compliance with the CO standards (ADEC 2001a). The health effects associated with exposure to CO were not addressed in these public-education campaigns. The committee is concerned that the public-education campaign has not sufficiently emphasized the health concerns associated with high CO exposure and has thus reinforced to residents the notion that the federal standards are arbitrary.

Surveys are an important tool for assessing the effectiveness of public-education activities and identifying avenues for improved efforts. The borough conducted a survey in November 2000 that indicated that its public-education activities may have been successful (ADEC 2001a). The use of public transportation has increased, and the awareness of plugging in at temperatures above 0°F seems to have increased. However, 86% of people polled listed improved vehicle starting as one of the major reasons for plugging in and only 54-56% listed improved air quality and public health as a major reason. Education and marketing have played important roles and will probably continue to do so in the borough's efforts to attain the NAAQS for CO. Given the large potential outlays of money in public-education campaigns, the committee recommends that studies be undertaken to assess their effectiveness.

> **BOX 2-8** Recommendations: Public Education
>
> Public-education programs should be continued and expanded to increase public awareness of the potential health effects of high ambient CO concentrations and to increase public participation in efforts to improve air quality. Surveys of public opinion should be used in designing the programs and assessing their effectiveness.

Methods and Data to Quantify Control-Strategy Effects

Evaluation of CO-control strategies is important for two reasons: to determine whether an intervention strategy is working and is cost-effective and to identify ways to improve it. Because of its importance as a local emissions-control strategy, I/M evaluation is recommended by the committee, as discussed earlier. Other elements of the borough's CO-control strategy would also benefit from evaluation. Ideally, such evaluations should include the following three components:

- Adoption: the extent to which the intervention is delivered successfully to the target audience; for example, the number of vehicles using electric plug-ins or engine heaters because of a public-education and marketing program that promotes broader use of plug-ins to reduce cold-start CO emissions or the rate of compliance with I/M requirements.
- Efficacy: the potential effect of the intervention on the intended outcome when the intervention is delivered successfully; for example, the reduction in the total CO emissions attributable to each vehicle that uses an electric plug-in engine heater or the reduction in emissions attributable to each vehicle that completes the I/M program (including the ability of the I/M program to identify high-emitting vehicles, the repair efficiency for those that are repaired, and removal of those not repaired).
- Effectiveness: the overall effect of the intervention program on the intended outcome; for example, the reduction in CO emissions per day attributable to the public-education program promoting plug-ins (the "bottom line" of the intervention program) or the overall emissions reduction attributable to the I/M program.

The overall program effectiveness depends on both adoption and efficacy. If plug-ins did not reduce CO emissions from individual vehicles (were not efficacious), the promotion program would be ineffective even if it were successful in increasing the use of plug-ins broadly. But an intervention that is highly efficacious but is not adopted widely is also ineffective. The efficacy of strategies that directly affect emissions from vehicles can be estimated with laboratory and field experiments, such as the tests that show emissions reductions that occur when high-emitting vehicles are repaired. Program evaluation attempts to synthesize data from various sources to combine efficacy and adoption and estimate overall program effectiveness. For example, the efficacy of repairs of high-emitting vehicles in a laboratory setting needs to be evaluated more extensively in repair shops and combined with data on compliance and other factors to evaluate the overall effectiveness of a full I/M program.

Generally, the assessment of program adoption and overall program effectiveness requires causal inference to determine the portions of the adoption and of the overall outcome that are attributable to the intervention program. For example, the borough may want to determine how much of the increase in bus ridership is attributable to having free bus service during the winter months, that is, how many of the riders would have driven a private vehicle if the incentive were not provided. The ideal method for developing causal inference is a randomized trial. In evaluating the ridership effect, a pilot study to determine consumer receptivity of free bus service might randomize and track samples of residents who do and do not receive a waiver of bus fare. When randomized trials are not possible due to resource limitations and other factors, such as sampling selection bias, a pre-post design is common in program evaluation. With pre-post methods, researchers assess program adoption or outcomes before and after the implementation of the intervention program and then examine the changes to assess their effect on program effectiveness. For example, I/M programs can be evaluated by comparing vehicles that have undergone emissions testing in an area with vehicles that have not or by comparing vehicle emissions in an area that has emissions testing with emissions in an area that does not. However, it is often challenging to use pre-post methods to determine whether an observed change is due to the intervention program or to changes in other factors.

Control-Strategy Benefits for Related Pollutants

Reducing the emission of CO during cold starts would have additional benefits, because other pollutants are also formed during fuel-rich conditions. The amounts of HC, particulate matter (PM), and other toxic emissions are a strong function of air-to-fuel ratio and therefore correlate strongly with CO emissions under cold-start conditions. The emissions of such pollutants as benzene, 1,3-butadiene, polycyclic aromatic hydrocarbons (PAHs), and aldehydes are strongly favored during fuel-rich operation (SAE 1992).

Hydrocarbon (HC) emissions are largely a result of the excess liquid fuel injected during a cold start. The lower the engine temperature, the larger the total amount of fuel called for by the engine control unit to ensure adequate starting. Heating the engine block or coolant reduces the total requirement for fuel injection and has a direct effect on the emission of unburned fuel during a cold start. Because benzene is present in small amounts in the fuel, benzene emissions result from evaporation and emission of unburned benzene as well as from incomplete combustion of other gasoline components. Aldehydes, PAHs, and PM generated during partial oxidation increase during fuel-rich operation. Therefore, methods that reduce the amount of fuel injected in the engine or increase oxidation in the catalyst reduce the amounts of fuel-derived pollutants. However, strategies for reducing CO emissions will not always reduce all HC emissions. The use of oxygenated fuels for control of CO should decrease the concentrations of PAHs and benzene but would most likely increase emissions of aldehydes (which one depends on the oxygenate) (SAE 1992). Likewise, switching from gasoline to diesel-fueled vehicles will reduce CO but may be associated with higher emissions of NO_x and PM.

PM emissions from gasoline engines are also sensitive to ambient temperatures. A study performed at the Alaska Department of Environmental Protection (Mulawa et al. 1997) showed a linear correlation of PM_{10} with HC and CO emissions. Therefore, CO emissions-reduction measures would probably also have a beneficial effect on the emissions of correlated pollutants, such as HC, PM_{10} and $PM_{2.5}$, and PAHs.

Mechanical Vertical Mixing

The combination of the extremely shallow depth and strong stability of the atmospheric boundary layer in Fairbanks and the possibly small spatial extent of high CO concentrations provides an opportunity to consider mechanical means to disperse CO. Wind turbines that destabilize the boundary layer are used in agricultural areas to mix cold air upward and warm air downward to prevent crops from freezing on calm, clear nights. Those types of turbines might also be used in the borough to mix CO-laden air upward and clean air downward. The objective is not to blow CO out of the nonattainment area but to mix and dilute the air vertically; this is energetically more viable.

Figure 2-10 illustrates the depth and extent of mixing that such wind machines can achieve. The figure shows the vertical cross section of temperature responses produced by a 320-lb thrust wind turbine. As shown in this figure, the turbine-induced vertical mixing brings warmer air downward. Stippled areas represent trees and solid vertical lines indicate masts on which temperature sensors were mounted. Crawford (1965) indicated that one machine with a thrust of 1,000 lb can influence an area of 20 acres and that the most effective mixing of large areas is with slowly rotating turbines. Bates (1972) reported that two turbines, each with a thrust of 1,310 lb and rotating once every 9 min, could warm an area of 23 acres substantially.

The feasibility and the various local and regional effects of a mechanical mixing approach would need to be thoroughly researched before implementation. To explore the concept further, the borough could pursue physical or numerical modeling. For example, the Fairbanks central urban area could be represented in a wind-tunnel model with a stable thermal stratification designed to mimic that of Fairbanks. Pielke (2002) illustrates the appropriateness of using such physical modeling in stably stratified locations. CO, or a

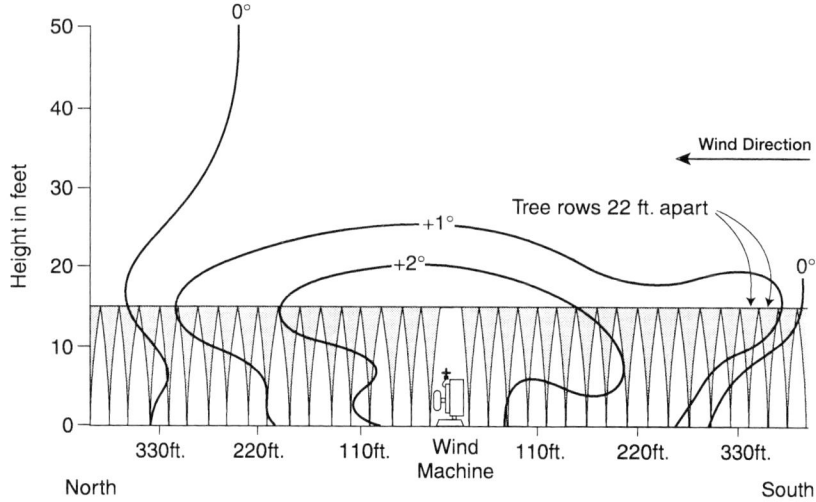

FIGURE 2-10 Illustration of the depth and extent of mixing that wind machines can achieve. Solid lines indicate temperature change (in degrees Fahrenheit) due to induced mixing. Source: Crawford 1965. Reprinted with permission from Meteorological Monographs; copyright 1965, American Meteorological Society.

surrogate gas, could be input at the lowest level of the physical model. Alternatively, the spatial scale of the downtown area is small enough to use detailed fluid-dynamics models (such as large-eddy simulation) to conduct numerical experiments on the effect of wind turbines in mixing air and the effect of buildings in dissipating air movement.

As discussed previously, the spatial extent of the high CO concentrations found in Fairbanks needs to be known to evaluate the value of such an approach. Furthermore, cost, noise, and safety considerations could be important in such a strategy and must be assessed. Because of these concerns, the committee feels that other emissions-control strategies should be pursued before further investigating a strategy of mechanical vertical mixing.

SIMPLE BOX MODEL OF THE BOROUGH

Various models that have been used for attainment demonstrations, research purposes, and air quality forecasting were discussed in Chapter 1. In its most recent SIP, Fairbanks used a statistical rollback model to estimate the emissions reductions needed to reach attainment (ADEC 2001a). Although more sophisticated tools (such as dispersion and urban-airshed models) are available, Fairbanks claimed in its SIP that it did not have sufficient information about its meteorology or emissions inventories to use them. Nevertheless, the borough acknowledges the limitations of the rollback approach and is committed to pursuing the development of dispersion modeling capabilities. In addition to improving the meteorological data and emissions inventory, the borough will also have to address the limitations of widely available dispersion models under conditions of severe temperature inversions and very low windspeeds (Bowling 1985).

In the meantime, the committee has developed a simple box model for the borough nonattainment area to gain some insight into the roles played by emissions and meteorological variables. This type of model can also be used to test various emissions scenarios and possibly for attainment demonstration. The model solves for the time-varying CO concentration as a function of CO emissions and meteorological conditions in the nonattainment area, about 88 km^2—6.4 km north-south by 13.8 km east-west, centered on downtown Fairbanks. Uniform conditions are assumed for the volume of the box. Specifically, CO concentrations are assumed to be well mixed, and meteorological variables—windspeed and wind direction, temperature, pressure, and mixing-layer height—are constant throughout the box. It is known from observations that those quantities vary over the nonattainment area and probably have a substantial effect on local CO concentrations, but the variations are not resolved here. In addition, it is assumed that there is negligible CO in the air entering the box—a reasonable assumption for a box of this size and for a city far from other CO sources.

To demonstrate how a simple box model works, the results are presented for an average 72-h CO event in which the 9-ppm standard is exceeded during the middle day (hours 24-48). The event was calculated by averaging hourly observations of CO concentrations, vehicle counts, and meteorological variables during five of the last six exceedances in Fairbanks. The five events averaged were centered on 2/23/98, 2/11/99, 2/16/99, 11/19/99, and 2/8/00; an exceedance of the 8-h CO standard also occurred on 2/24/98. The highest 8-h average was observed four of the five times at the Post Office and once (2/11/99) at the Hunter School. A more rigorous application of the model would be to simulate a single exceedance.

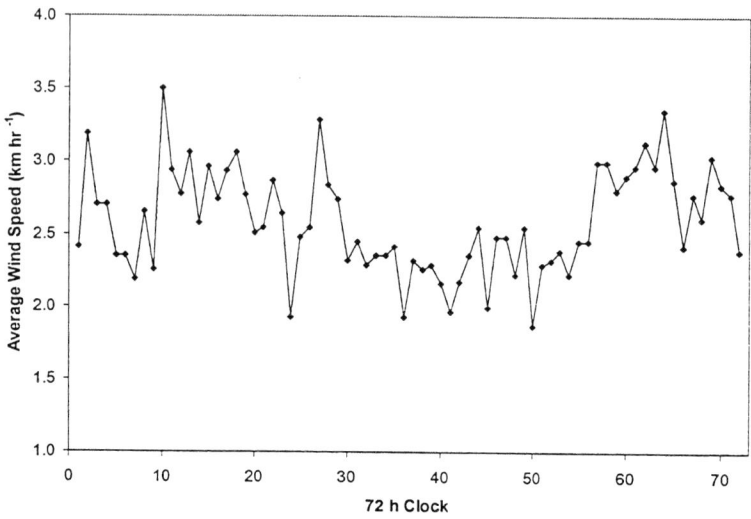

FIGURE 2-11 Windspeed (km/h) measured at 10 m height at the meteorological tower in downtown Fairbanks. The hourly values are averaged over the five exceedance events.

The Data

The model inputs include average hourly values of meteorological variables and vehicle counts and average daily emissions rates. Temperature, pressure, and windspeed (Figure 2-11) were measured at a 10-m altitude at the meteorological tower in downtown Fairbanks.[12] The temperature-inversion strength ($-\Delta T/\Delta z$) is computed from temperature measurements at 3 and 10 m (Figure 2-12). The air density is calculated from the hourly average temperature and pressure, assuming that air is an ideal gas. Variations in these meteorological parameters might be much larger for the original measurements than for the averages shown here. Emissions of CO in the box are assumed to be 8.4 tpd from stationary sources and 16.5 tpd from mobile sources, fol-

---

[12]The CO, meteorological, and vehicle-count data were kindly supplied by Paul Rossow of the borough Air Quality Office and Paul Prusak of the Alaska State Department of Transportation.

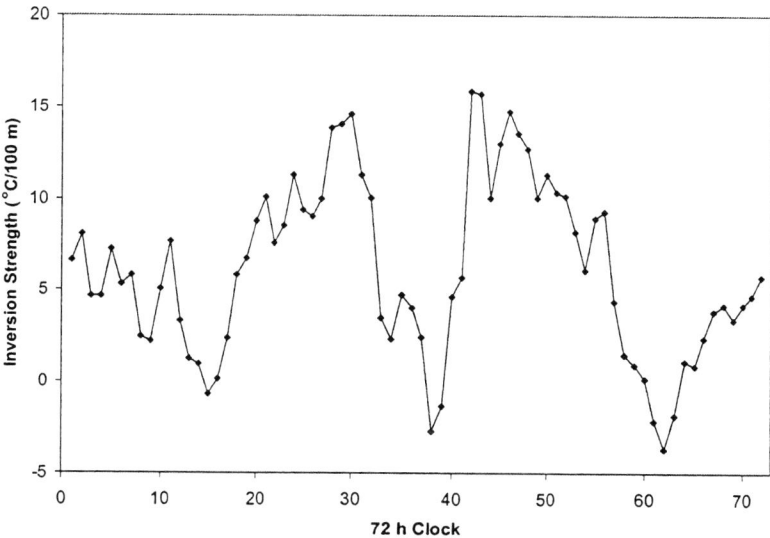

FIGURE 2-12 Inversion strength (°C/100 m) computed as the temperature difference between 3 and 10 m heights measured at the meteorological tower in downtown Fairbanks. The hourly values are averaged over the five exceedance events.

lowing the values estimated by Sierra Research in a 1999 CO emissions inventory.[13] Mobile sources in that inventory are estimated to be 7.0 tpd from cold-start and initial-idle emissions and 9.5 tpd from traveling emissions. Observed vehicle counts in downtown Fairbanks (at locations shown in Figure 2-4) are used to convert the daily mobile-source emissions rate to hourly emissions (Figure 2-13). The model performance is evaluated on the basis of CO measurements made at the three monitoring sites in downtown Fairbanks. The CO observations are averaged hourly and over the three monitoring sites.

The Model

The mass of CO in a box representing the nonattainment area can be determined with a mass-balance expression (assuming that CO concentrations

---

[13]Data provided by Robert Dulla, Sierra Research.

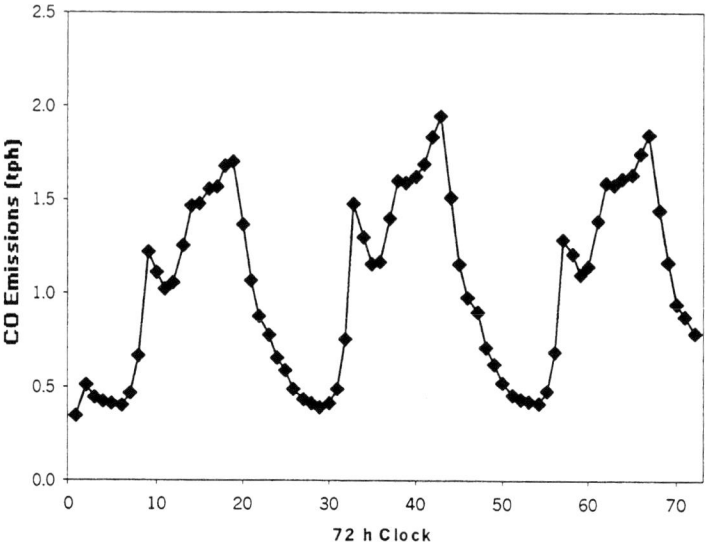

FIGURE 2-13 CO emissions ($E[t]$) in tons per hour (tph) for the average exceedance event.

outside the box are negligible):

$$\frac{dm(t)}{dt} = E(t) - \frac{m(t)S(t)}{L}, \qquad (1)$$

where

t = time (h);
$m(t)$ = mass of CO in box (ton);
$E(t)$ = emissions of CO in box (tph);
$S(t)$ = windspeed (km/h); and
L = length of box (km) in direction of wind.

Using the relationship between mass and concentration:

$$m = c(t)\rho(t)V(t), \qquad (2)$$

where

$c(t)$ = concentration (ppm);

Fairbanks Case Study 103

$\rho(t)$ = air density in the box (g/km³); and
$V(t)$ = volume of the box (km³).

The mass-balance equation can be expressed in units of concentration:

$$\frac{dc(t)}{dt} = \frac{E(t)}{\rho(t)V(t)} - \frac{c(t)S(t)}{L} - \frac{c(t)}{V(t)}\frac{dV(t)}{dt}\bigg|_{if\frac{dV(t)}{dt}>0}. \quad (3)$$

Finally, the concentration of CO at any time $c(t + \Delta t)$ can be estimated with Equation 3 as follows:

$$c(t + \Delta t) \approx c(t) + \frac{dc(t)}{dt}\Delta t$$

$$c(t + \Delta t) \approx c(t) + \left[\frac{E(t)}{\rho(t)V(t)} - \frac{c(t)S(t)}{L} - \frac{c(t)}{V(t)}\frac{dV(t)}{dt}\bigg|_{if\frac{dV(t)}{dt}>0}\right]\Delta t. \quad (4)$$

We solved Equation 4 by using a time increment, Δt, of 1 h.

Solving Equation 4 requires knowing or assuming time-dependent values of the variables on the right side. As discussed previously, observed values were available for windspeed, $S(t)$; air density, $\rho(t)$ (computed from temperature and pressure); emissions, $E(t)$; and an initial value for CO concentration, $c(t)$. The box length, L, was assumed to be 6.4 km, the length of the box from north to south, in the direction of the incoming wind. The major challenge was devising a function for the volume $V(t)$ that is reasonable and yields a calculated CO concentration that varies with time in a way similar to the observed hourly average. Because the area of the box is constant, the variations in the volume are based on variations in the height of the mixing layer, $H(t)$. The committee assumes that $H(t)$ depends on the observed inversion strength $\Delta T/\Delta z$) in the following way:

Stable conditions: $\Delta T/\Delta z < 0°C/100$ m $\qquad H = 8$ m

Mild inversion
conditions: $\qquad 0°C/100m < \Delta T/\Delta z < 11°C/100$ m $\qquad H = 5$ m

Severe inversion
conditions: $\Delta T/\Delta z > 11°C/100$ m $\qquad H = 2.25$ m

The mixing heights assigned that way are shown in Figure 2-14. They are highest in the early afternoon, when the atmosphere can become unstable (positive lapse rate, negative inversion) because of solar heating.

Mixing height during severe inversion conditions is determined by a combination of turbulent mixing (mostly from high-speed exhaust gas leaving vehicle tailpipes and vehicle motion under low-windspeed conditions) and convective mixing from the rising, initially hot exhaust gas. Thus, the mixing height is rather heterogeneous over the nonattainment area, and it is difficult to determine directly from observations a single height to use in the box model. For these reasons, the mixing heights used in the model were derived to best match the observed CO concentrations. As such, the model provides a useful tool for estimating an effective mixing height based on the constraints of estimated emissions and observed surface CO concentrations.

Results and Discussion

Concentrations of CO predicted by the model are compared with those of the average observed event in Figure 2-15. In general, the model does a good job of reproducing the magnitude of the CO concentration and the diurnal variation over the 72-h event. To simulate the high CO concentrations observed during the exceedance, the mixed-layer height needed to be low (2.25 m, about 7.4 ft) during severe inversions. Marginally less wind and more traffic on the average exceedance day (see Figures 2-11 and 2-13) did not yield the observed doubling of CO concentrations compared with the days before and after the exceedance. A box height of 2.25 m may be too low if the spatial variability in CO emissions invalidates the assumption of a well-mixed box. In particular, if CO emissions are greater in the downtown area (where the three CO monitors are) than elsewhere in the modeled area, the mixing height downtown could be higher.

The lowest CO concentrations (about 1 ppm) are observed in the early morning hours, between about 2 and 6 a.m. The model shows concentrations during those times about twice as high (about 2 ppm), suggesting that the stationary-source CO emissions used in the model (0.35 tph) are too high during these hours. The discrepancy could be due to insufficient horizontal dispersion in the model, but that does not appear to be the case. With an

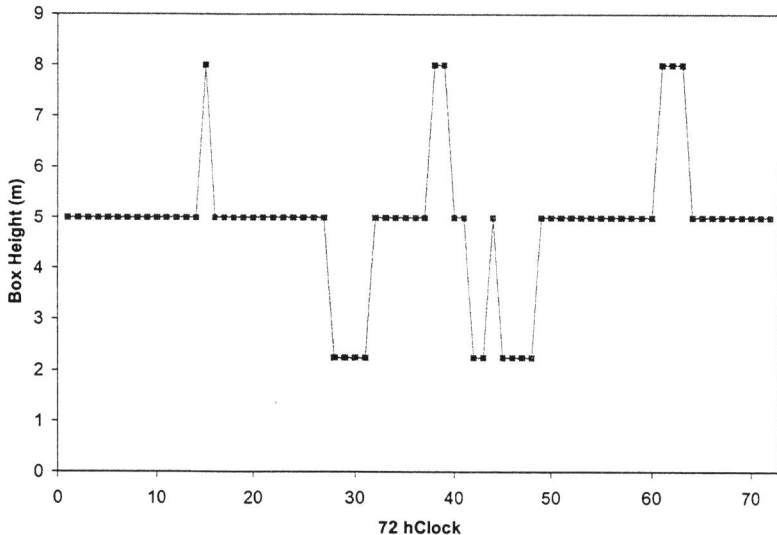

FIGURE 2-14 The box heights (m) used for the model fit shown in Figure 2-15.

assumed box length in the direction of the wind of 6.4 km and an average windspeed of 2.6 km/h, the time constant for horizontal dispersion in the model is about 2.5 h, certainly fast enough to deplete CO to low concentrations during the early morning hours, when emissions are low.

Although the model generally yields the correct shape of the diurnal variations of CO concentration, the calculated afternoon maximums lag those observed at about 5 p.m. The highest CO emissions in the model are between 5 and 6 p.m., when vehicle traffic is maximal. However, idling emissions in the downtown area before people leave work to go home have the effect of shifting the observed emissions maximum 30-60 min earlier than would be expected on the basis of vehicle counts. Improving the model requires an emissions inventory more detailed in time and space, particularly with respect to plug-in and idling behavior.

The box model described here provides an illustrative example of how this technique can be applied to better understand the roles of emissions and meteorology in high-CO episodes. The model could be adapted in a number of ways. As mentioned before, a specific exceedance event, rather than averaged events as shown here, could be analyzed. Another modification could be to consider a different surface area of the control volume. For example, the

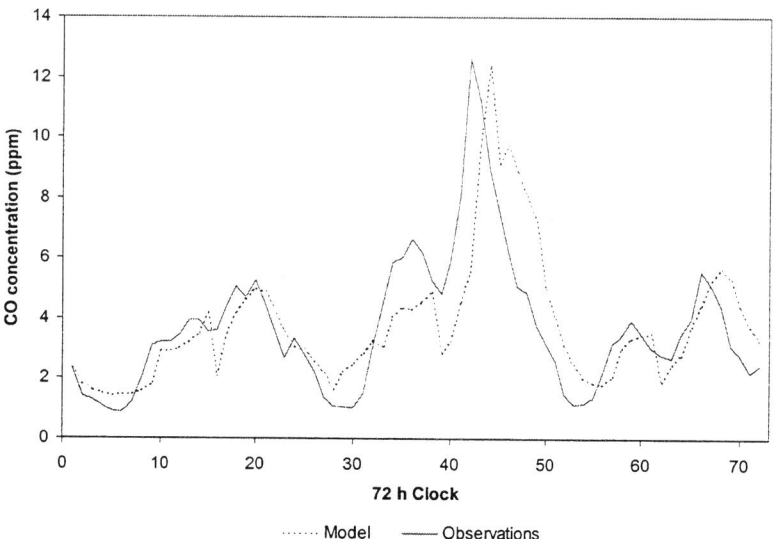

FIGURE 2-15 Comparison of simulated and observed CO concentrations (ppm) for the average exceedance event.

model could simulate only downtown Fairbanks. For this case, the assumption that no CO is transported into the box would need to be reassessed. A vertically resolved, one-dimensional model also could be used. That approach could help improve the understanding of how stationary sources affect surface CO concentrations.

In the short term, the committee recommends that the borough use a box-model approach for air quality planning and for demonstrating attainment. Unlike less physically comprehensive rollback models, a box model can provide information on how temporal changes in emissions and meteorology affect CO concentrations. Unlike more physically comprehensive airshed modeling, the monitoring data to implement a box model are readily available. The box-model approach used by the borough need not be identical to the one developed here; some of the modifications described above should be considered. The borough may want to use a box model in conjunction with a model that focuses on an area immediately surrounding a monitoring site, such as the probabilistic rollback models described in Chapter 1; however, the time and effort needed to develop inventories for these local modeling approaches may be better spent developing more physically comprehensive regional modeling capabilities for the Fairbanks area. For the future, the committee recommends that the data needed to initiate and validate airshed modeling be gathered and that more physically comprehensive models be used.

> **BOX 2-9** Recommendations: Improving Ambient-CO Modeling in the Borough
>
> In the near term, Alaska should use a simple box-model approach, which simulates the effects of emissions and meteorology in a well-mixed control volume, for air quality planning purposes in Fairbanks. Such an approach could provide greater insights into the effects of the timing of CO emissions and meteorological variables. The relative contributions of mobile and stationary sources to CO episodes could also be assessed with this type of model. Enhanced data-collection efforts are required to support this and more sophisticated modeling efforts.
>
> Improvements in the statistical forecasting approach used by the borough might help in forecasting episodes of high CO concentrations. More work is also needed to develop, apply, and evaluate more sophisticated, physically comprehensive models that would simulate how CO concentrations vary with time and space over the entire borough. Such models could be used for planning, forecasting, and assessing human exposure to high CO concentrations. It is important that model development and testing be specific to the extreme conditions that occur in Fairbanks. Model development must occur in concert with improved monitoring to enable model evaluation.

SUMMARY

Fairbanks presents a challenge for air quality management. It constitutes an extreme example of the roles of topography and meteorology in producing air quality problems. In winter, the area is subject to extreme ground-level inversions. Although no industries in the region emit large amounts of CO or other pollutants, the inversions are extremely effective in trapping the products of incomplete combustion that are emitted near ground level, particularly CO from vehicles. Monitors in the downtown area have measured as many as 130 d/y with exceedances of the 8-h CO standard. The situation has greatly improved over the last 30 y. No exceedances have occurred over the last 2 y.

The days currently most susceptible to high CO concentrations in Fairbanks are not the coldest days. Meteorological conditions in Fairbanks lead to exceedances of the CO health standard primarily when the ambient temperatures typically are -20 to 20°F. The combination of human behavior and motor-vehicle technology further narrows the primary temperatures of concern to 0-20°F. Air quality planning and controls should focus on such days. The committee has identified a number of options available to the borough and Alaska to reduce CO emissions further. Among them, improving the vehicle I/M program has high priority, as does an enhanced plug-in program. Use of low-sulfur fuel would also help. An enhanced alert-day program might provide needed reductions on critical days. And there is a need to inform and educate the community better about the health effects of exposure to air pollutants and ways to improve air quality.

Improved monitoring and characterization of CO concentrations in the area are needed. Modeling CO in the borough is a serious challenge, but it can be helpful in planning and in forecasting possible exceedance days. For now, a simple box model is suggested for planning purposes; more sophisticated models either are inappropriate for the conditions or require much more extensive monitoring data. Statistical models can be used, for now, to help in forecasting, but work is needed to improve them. More research with comprehensive models is needed for future application for both forecasting and planning. These more comprehensive models will also require improvements in the emissions inventory, including the nonroad, area, and point sources that contribute to high-CO episodes.

3

Implications of the Fairbanks Case Study

PROSPECTS FOR CONTINUED ATTAINMENT

Ambient carbon monoxide (CO) concentrations in Fairbanks are the result of human activities, especially the use of motor vehicles, and of regional and local meteorology and topography. Severe temperature inversions and low windspeeds, prevalent in Fairbanks during winter, trap CO emitted close to the ground. Although the number of violations of the 8-h National Ambient Air Quality Standard (NAAQS) for CO has declined substantially over the past 25-30 y, the inversion conditions experienced in Fairbanks are among the most severe in the United States. The prospects for continued attainment in Fairbanks depend essentially upon whether the current understanding of the cause-effect relationships at work in the region is accurate, whether observed trends in human activity and emissions continue as expected, and whether meteorological conditions are favorable for CO dispersal.

As stated earlier, maintaining compliance with the NAAQS for CO will be unlikely without an accurate understanding of cause-effect relationships and will depend on how this understanding is translated into control policies. The Fairbanks North Star Borough and Alaska have invested more resources in characterizing the cause-effect relationships involved in Fairbanks's CO problem than have other cities of similar size in the United States. Management of CO in the borough has focused on control strategies that reduce overall emissions from motor vehicles (primarily through the I/M program) and that reduce cold-start emissions (primarily through vehicle plug-ins). Analysis has focused on how the coincidence of cold starts and the strengthening of inversion conditions during winter afternoons can play a major role in CO exceedances. However, little attention has been paid to characterizing other sources of CO, which may be highly uncertain. The spatial extent of the CO problem, and the understanding of the microscale and regional meteorological conditions that are associated with exceedances, also are poorly characterized. Data on emissions from vehicles operating under winter conditions in Fairbanks are sparse, so it is difficult to predict the effects of future vehicle emissions-control strategies.

The borough recently has experienced decreasing emissions from motor vehicles, and this contributes to a higher likelihood of attaining and maintaining compliance. Federal controls on vehicles and fuels have had by far the largest effects in reducing CO emissions. Most of the recent reductions have been attributed to more stringent new-vehicle certification standards for CO. State and local programs generally help to reduce emissions but do not have the same impact. Nonetheless, in the absence of further federal mandates designed to yield additional emissions reductions in cold climates, enhancement of state or local controls is essential for achieving and maintaining CO concentration standards.

Long-term population growth in the area remains an issue. It is likely that a natural-gas pipeline or missile-defense initiative in Alaska will be approved and funded. If either of those activities occur, the growth in population, in vehicle-miles traveled, in service-industry activity, and in construction activities would substantially increase the probability of a CO-attainment lapse in Fairbanks in the future.

Finally, the prospects for continued attainment in Fairbanks depend on meteorological conditions. Severe inversions are inevitable. Exceedances could be avoided if such inversions occurred on days when emissions-producing activities were at a minimum or if emissions were minimized on days with severe inversions. If researchers could reliably discern the meteorological patterns that cause severe inversions, short-term mitigation strategies could be developed and used. But there is insufficient knowledge to understand the microclimate of the Fairbanks basin, let alone to understand how large-scale weather patterns in Alaska influence inversion conditions. Without improved meteorological understanding, the region cannot know whether additional emissions-reduction efforts or other mitigation strategies will be necessary until it is too late to prevent the area from slipping back into nonattainment.

The committee concludes that Fairbanks will be susceptible to violating the CO health standards for many years because of its severe meteorological conditions. That point is underscored by a December 2001 exceedance of the standard in Anchorage, which had no violations over the last 3 y. A modest change in regional or local meteorology or unforeseen growth in vehicle activity or emissions could readily cause Fairbanks to violate the standard again. Sustained efforts will be required if Fairbanks is to maintain compliance with the standard over the long term.

Nonetheless, Fairbanks has made great progress in reducing its violations of the 8-h CO standard. The success in reducing the number of days of violations from over 130 during 1973-1974 to below 4 per year over the last 5 y (and none in the last 2 y) demonstrates the ability of new vehicle-emissions standards and local controls to reduce motor-vehicle emissions. That improvement has reduced population exposures to CO and related pollutants, but some probably remain.

RELATIONSHIP OF FAIRBANKS CASE STUDY TO OTHER NONATTAINMENT AREAS

A number of other areas in the United States are struggling to attain the CO NAAQS, but Fairbanks is unique in the severity of its inversions. The Fairbanks area is subject to strongly inhibited mixing in winter because it has little solar heating, which prevents deepening of the mixing layer during daytime, and very light winds, which result in slow exchange of air between the city and its surroundings. Such an environment is conducive to the accumulation of pollution released near the ground. Nonattainment areas farther south have similar conditions in winter but benefit from additional sunlight that allows more vertical turbulent mixing. Although such mixing may not occur on cloudy days, these conditions rarely persist for long periods in association with light or calm winds. Even in mountain valleys, where terrain can be effective in trapping pollution, the solar heating of the surface and the relatively frequent windy conditions provide better dilution of pollutants than that in the Fairbanks area.

Other areas in nonattainment for CO must address unique conditions that contribute to their own CO problems. For example, Calexico, California, is close to the U.S.-Mexico border, where pollution from vehicles waiting to cross the border is thought to be the primary cause of high CO concentrations. Las Vegas, Nevada, not only has problematic meteorological and topographical conditions but is also one of the fastest-growing areas in the United States. Lynwood, California, has both severe inversion conditions and higher than average vehicle emissions that contribute to CO violations.

In many areas of the Unites States that have air quality problems related to CO and other pollutants, meteorological or topographical conditions exacerbate the accumulation of pollution. But health benefits can still be gained by meeting the health standards set by EPA. Although the conditions in Fairbanks are severe, Fairbanks is in no way unique in having natural conditions that contribute substantially to its air quality problem.

References

ADEC (Alaska Department of Environmental Conservation). 2001a. State Air Quality Control Plan, Vol. 2. Analysis of Problems, Control Actions, Section III.C: Fairbanks Transportation Control Program. Adapted July 27, 2001. ADEC, Juneau, AK.

ADEC (Alaska Department of Environmental Conservation). 2001b. Amendments to State Air Quality Control Plan, Vol. 3. Appendices (to Volume II, Section III). Adapted July 27, 2001. ADEC, Juneau, AK.

ADEC (Alaska Department of Environmental Conservation). 2001c. State Air Quality Control Plan, Vol. 2. Analysis of Problems, Control Actions, Section III.B: Anchorage Transportation Control Program. Public Review Draft. August 30, 2001. ADEC, Juneau, AK.

Akland, G.G., T.D. Hartwell, T.R. Johnson, and R.W. Whitmore. 1985. Measuring human exposure to carbon monoxide in Washington, D.C., and Denver, Colorado, during the winter of 1982-1983. Environ. Sci. Technol. 19(10):911-918.

Allred, E.N., E.R. Bleecker, B.R. Chaitman, T.E. Dahms, S.O. Gottlieb, J.D. Hackney, M. Pagano, R.H. Selvester, S.M. Walden, and J. Warren. 1989a. Short-term effects of carbon monoxide exposure on the exercise performance of subjects with coronary artery disease. N. Engl. J. Med. 321(21):1426-1432.

Allred, E.N., E.R. Bleecker, B.R. Chaitman, T.E. Dahms, S.O. Gottlieb, J.D. Hackney, D. Hayes, M. Pagano, R.H. Selvester, S.M. Walden, et al. 1989b. Acute effects of carbon monoxide exposure on individuals with coronary artery disease. Res. Rep. Health Eff. Inst. (25):1-79.

Bates, E.M. 1972. Temperature inversion and freeze protection by wind machine. Agric. Meteorol. 9:335-346.

Beard, R.R., and G.A. Wertheim. 1967. Behavioral impairment associated with small doses of carbon monoxide. Am. J. Public Health Nations Health. 57(11):2012-2022.

Beckett, W.S. 1994. The epidemiology of occupational asthma. Eur. Respir. J. 7(1):161-164.

Benignus, V.A., D.A. Otto, J.D. Prah, and G. Benignus. 1977. Lack of effects of carbon monoxide on human vigilance. Percept. Mot. Skills 45(3 Pt.1):1007-1014.

Benson, J.D., V. Burns, R.A. Gorse, A.M. Hochhauser, W.J. Koehl, L.J. Painter, and R.M. Reuter. 1991. Effects of Gasoline Sulfur Level on Mass Exhaust Emissions—Auto/Oil Air Quality Improvement Research Program. SAE 912323. Warrendale, PA: Society of Automotive Engineers.

Bishop, G.A., D.H. Stedman, M. Hektner, and J.D. Ray. 1999. An in-use snowmobile emission survey in Yellowstone National Park. Environ. Sci. Technol. 33(21):3924-3926.

Bowen, J.L., P.A. Walsh, and R.C. Henry. 1996. Analysis od Data from Lynwood Carbon Monoxide Study. ARB-R-97/630. PB97-148779. Sacramento, CA: California Air Resources Board, Research Division.

Bowling, S.A. 1984. Meteorological Factors Responsible for High CO (Carbon Monoxide) Levels in Alaskan Cities. Final Report. EPA 600/3-84-096. NTIS PB85-115137. U.S. Environmental Protection Agency, Washington, DC. October 1984.

Bowling, S.A. 1985. Modifications necessary to use standard dispersion models at high latitudes. Atmos. Environ. 19(1):93-97.

Bowling, S.A. 1986. Climatology of high-latitude air pollution as illustrated by Fairbanks and Anchorage, Alaska. J. Clim. Appl. Meteorol. 25:22-34.

Bowling, S.A., and C.S. Benson. 1978. Study of the Subarctic Heat Island at Fairbanks, Alaska. EPA 600/4-78-027. Environmental Sciences Research Laboratory, Office of Research and Development, U.S. Environmental Protection Agency, Research Triangle Park, NC. 150 pp.

Boycott, W., and J. Cherry. 2001. Gasoline Manufacturing and Formulation in Alaska. Presentation to the Committee on Carbon Monoxide Episodes in Meteorological and Topographical Problem Areas, National Research Council, Fairbanks. AK, August 9, 2001.

Burnett, R.T., R.E. Dales, J.R. Brook, M.E. Raizenne, and D. Krewski. 1997. Association between ambient carbon monoxide levels and hospitalizations for congestive heart failure in the elderly in 10 Canadian cities. Epidemiology 8(2):162-167.

CARB (California Air Resources Board). 2000. Evaluation of California's Enhanced Vehicle Inspection and Maintenance Program (Smog Check II). Draft. California Air Resources Board, Sacramento, CA.

Cadle, S.H., P.A. Mulawa, E.C. Hunsanger, K. Nelson, R.A. Ragazzi, R. Barrett, G.L. Gallagher, D.R. Lawson, K.T. Knapp, and R. Snow. 1999. Composition of light-duty motor vehicle exhaust particulate matter in the Denver, Colorado area. Environ. Sci. Technol. 33(14):2328-2339.

Coburn, R.F. 1970. Enhancement by phenobarbital and diphenylhydantoin of carbon monoxide production in normal man. N. Engl. J. Med. 283(10):512-515.

Colorado Department of Public Health and Environment. 2000. Carbon Monoxide Redesignation Request and Maintenance Plan for the Denver Metropolitan Area. Colorado Department of Public Health and Environment, Air Pollution Control Division, Denver, CO. [Online]. Available:http://www.cdphe.state.co.us/ap/down/sipdenco.pdf [January 28, 2002].

Crawford, T.V. 1965. Frost protection with wind machines and heaters. Meteor. Monogr. 6(28):81-87.

Darlington, T.L., and D.F. Kahlbaum. 1998. Development of an Updated Cold CO Model. Report No. 032198. Prepared for the American Automobile Manufacturers Association by Air Improvement Resource, Inc, Novi, MI. March 21, 1998.

Davis, S.C. 1997. Transportation Energy Data Book: Edition 17. ORNL-6919. Center for Transportation Analysis, Oak Ridge National Laboratory, Oak Ridge, TN.

EPA (U.S. Environmental Protection Agency). 1992. Guideline for Modeling Carbon Monoxide from Roadway Intersections. EPA 454/R-92-005. NTIS PB 93-210391. Office of Air Quality Planning and Standards, U.S. Environmental Protection Agency, Research Triangle Park, NC.

EPA (U.S. Environmental Protection Agency). 1993. Clean Cars for Clean Air: Inspection and Maintenance Programs. Fact Sheet OMS-14. EPA 400-F-92-016. Office of Mobile Sources. U.S. Environmental Protection Agency. Ann Arbor, MI. [Online]. Available: http://epa.gov/otaq/14-insp.htm. [February 16, 2001].

EPA (U.S. Environmental Protection Agency). 1994. EPA Health Effects Notebook for Hazardous Air Pollutants-Draft. EPA-452/D-95-003. Air Risk Information Support Center, U.S. Environmental Protection Agency, Research Triangle Park, NC. [Online]. Available:http://www.epa.gov/ttn/atw/hapindex.html [March 25, 2002].

EPA (U.S. Environmental Protection Agency). 1997. National Ambient Air Quality Standards for Particulate Matter; Final Rule. Fed. Regist. 62(138):38651-38760. (July 18, 1997).

EPA (U.S. Environmental Protection Agency). 1998a. AP-42, Air Pollutant Emission Factors, 1998. AP-42. Office of Mobile Sources, U.S. Environmental Protection Agency. [Online]. Available: http://www.epa.gov/ttn/chief/ap42/ [January 29, 2002].

EPA (U.S. Environmental Protection Agency). 1998b. Inspection and Maintenance (I/M) Program Effectiveness Methodologies. EPA 420-S-98-015. Office of Air and Radiation. U.S. Environmental Protection Agency, Ann Arbor, MI.

EPA (U.S. Environmental Protection Agency). 2000a. Air Quality Criteria for Carbon Monoxide. EPA/600/P-99/001F. Office of Research and Development, U.S. Environmental Protection Agency, Research Triangle Park, NC. [Online]. Available: http://www.epa.gov/NCEA/co/index.html [January 28, 2002].

EPA (U.S. Environmental Protection Agency). 2000b. 1998-2000 Ozone and 1999-2000 Carbon Monoxide. Air Quality Update. Office of Air and Radiation, U.S.

Environmental Protection Agency, Washington, DC. [Online]. Available: http://www.epa.gov/oar/aqtrnd00/carboz00.html [January 28, 2002].

EPA (U.S. Environmental Protection Agency). 2000c. Air pollution control; new motor and engines: Tier 2 motor vehicle emission standards and gasoline sulfur control requirements. Fed. Regist. 65(28):6697-6870. (February 10, 2000).

EPA (U.S. Environmental Protection Agency). 2001a. National Air Quality and Emissions Trends Report, 1999. EPA/454/R-01-004. Office of Air Quality Planning and Standards, Emissions Monitoring and Analysis Division, Air Quality Trends Analysis Group, U.S. Environmental Protection Agency, Research Triangle Park, NC. March 2001. [Online]. Available: http://www.epa.gov/oar/aqtrnd99/ [January 28, 2002].

EPA (U.S. Environmental Protection Agency). 2001b. Air Quality Criteria for Particulate Matter, Vol. 1. and Vol. 2. EPA/600/P-99/002aB. EPA/600/P-99/002bB. (Second External Review Draft). Office of Research and Development, U.S. Environmental Protection Agency, Washington, DC. March 2001.[Online]. Available: http://cfpub.epa.gov/ncea/cfm/partmatt.cfm?ActType=default [March 25, 2002].

EPA (U.S. Environmental Protection Agency). 2001c. The Projection of Mobile Source Air Toxics from 1996 to 2007: Emissions and Concentrations. EPA/420/R-01-038. (Draft Report). U.S. Environmental Protection Agency, Research Triangle Park, NC. August 2001. [Online]. Available: http://www.epa.gov/otaq/toxics.htm#regs [March 25, 2002].

EPA (U.S. Environmental Protection Agency). 2001d. Draft Guidance on the Use of Remote Sensing for Evaluation of I/M Program Performance. EPA420-P-01-002. Certification and Compliance Division, Office of Transportation and Air Quality. U.S. Environmental Protection Agency. July 2001. [Online]. Available: http://www.epa.gov/otaq/ epg/progeval.htm [January 28, 2002].

EPA (U.S. Environmental Protection Agency). 2001e. Draft Guidance on Use of In-Program Data for Evaluation of I/M Program Performance. EPA420-P-01-003. Certification and Compliance Division, Office of Transportation and Air Quality. U.S. Environmental Protection Agency. August 2001. [Online]. Available: http://www.epa.gov/otaq/ epg/progeval.htm [January 28, 2002].

EPA (U.S. Environmental Protection Agency). 2001f. TCM Programs Organized by Program Category. Transportation Control Measures Program Information Directory, U.S. Environmental Protection Agency. [Online]. Available: http://yosemite.epa.gov/aa/tcmsitei.nsf/72c5a95b733f6827852567450074a45a?OpenView [January 29, 2002].

EPA (U.S. Environmental Protection Agency). 2002. Databases. Transportation Air Quality Center. Office of Transportation and Air Quality (OTAQ), Office of Air and Radiation, U.S. Environmental Protection Agency. [Online]. Available: http://www.epa.gov/ otaq/transp/ [March 25, 2002].

References 117

FHWA (Federal Highway Administration). 2000.. Traffic Software Integrated System (TSIS) Version 5.0. Final. Federal Highway Administration. [Online]. Available: http://www.fhwa-tsis.com/ [March 25, 2002].

Flachsbart, P.G., G.A. Mack, J.E. Howes, and C.E. Rodes. 1987. Carbon monoxide exposures of Washington commuters. JAPCA 37(2):135-142.

Guay, G. 2001. Fairbanks Carbon Monoxide Study 1999-2001. State of Alaska, Department of Environmental Conservation, Division of Air and Water Quality. CD-ROM, Fairbanks CO Studies. Fairbanks, AK. October 31, 2001.

Guensler, R. 1989. Preparation of Alternative Emission Control Plans for Surface Coating Operations. California Air Resources Board, Sacramento, CA. Paper No. 89-118B.3. In proceedings: 82nd Annual Meeting and Exhibition of the Air & Waste Management Association, June 25-30, 1989, Anaheim, CA. Vol. 7. [Online]. Available: http://transaq.ce.gatech.edu/guensler/publications/proceedings/proceedings.html [March 25, 2002].

Guensler, R. 1998. Increasing Vehicle Occupancy in the United States. Pp. 127-155 in L'Avenir Des Déplacements en Ville (The Future of Urban Travel). O. Andan, et al., eds. Lyon, France: Laboratoire d'economie des transports. [Online]. Available: http://transaq.ce.gatech.edu/guensler/publications/books/books.html [March 25, 2002].

Hallmark, S.L., I. Fomunung, R. Guensler, and W. Bachman. 2001. Assessing impacts of improved signal timing as a transportation control measure using an activity-specific modeling approach. J.Transp. Res. Board 1738:49-55.

Harvey, G., and E. Deakin. 1993. A Manual of Regional Transportation Modeling Practice for Air Quality Analysis, Version 1.0. National Association of Regional Councils.

Heimrich, M.J. 1990. Air Injection to an Electrically-Heated Catalyst for Reducing Cold-Start Benzene Emissions from Gasoline Vehicles. SAE 902115. Warrendale, PA: Society of Automotive Engineers.

Heimrich, M.J., S. Albu, and J. Osborn. 1991. Electrically-Heated Catalyst System Conversions on Two Current-Technology Vehicles. SAE 910612. Warrendale, PA: Society of Automotive Engineers.

Henderson, G. 2001. Tesoro Alaska Gasoline Qualities and Specifications. Presentation to the Committee on Carbon Monoxide Episodes in Meteorological and Topographical Problem Areas, National Research Council, Fairbanks, AK, August 8, 2001.

Holmgren, B., L Spears, D. Wilson, and C. Benson. 1975. Acoustic soundings of the Fairbanks temperature inversions. Pp. 293-306 in Climate of the Arctic, G. Weller, and S.A. Bowling, eds. Fairbanks: Geophysical Institute, University of Alaska.

Hooke, W.H., and R.A. Pielke Jr. 2000. Short-term weather prediction: An orchestra in need of a conductor. Pp. 61-84 in Prediction: Science, Decision Making, and The Future of Nature, D. Sarewitz, R.A. Pielke Jr., and R. Byerly, eds. Washington, DC: Island Press.

Horvath, S.M., T.E. Dahms, and J.F. O'Hanlon. 1971. Carbon monoxide and human vigilance: A deleterious effect of present urban concentrations. Arch. Environ. Health 23(5):343-347.
Huschke, R.E., ed. 1959. Glossary of Meteorology. Boston: American Meteorological Society.
IMRC (California Inspection and Maintenance Review Committee). 2000. Smog Check II Evaluation. California Inspection and Maintenance Review Committee, Sacramento, CA. June 19, 2000. [Online]. Available: http://www.smogcheck.ca.gov/IMRC/ [January 30, 2002].
Klausmeier, R., and P. McClintock. 1998. The Greeley Remote Sensing Pilot Program—Final Report. Prepared for the Colorado Department of Public Health and Environment, by de la Torre Klausmeier Consulting, Inc., Austin, TX. and Applied Analysis. November 4, 1997.
Lyons, M. 2001. The Fairbanks North Star Borough's I/M Program and Public Transportation System. Presentation to the Committee on Carbon Monoxide Episodes in Meteorological and Topographical Problem Areas, National Research Council, Fairbanks, AK. August 8, 2001.
McFarland, R.A. 1970. The effects of exposure to small quantities of carbon monoxide on vision. Ann. N.Y. Acad. Sci. 174(1):301-312.
McFarland, R.A. 1973. Low level exposure to carbon monoxide and driving performance. Arch. Environ. Health 27(6):355-359.
Moolgavkar, S.H., E.G. Luebeck, and E.L. Anderson. 1997. Air pollution and hospital admissions for respiratory causes in Minneapolis-St. Paul and Birmingham. Epidemiology. 8(4):364-370.
Morris, R.D., E.N. Naumova, and R.L. Munasinghe. 1995. Ambient air pollution and hospitalization for congestive heart failure among elderly people in seven large U.S. cities. Am. J. Public Health 85(10):1361-1365.
Morris, S.S. 2001. CO in Anchorage: What have we learned in 20 years? Presentation to the NRC Committee on Carbon Monoxide Episodes in Meteorological and Topographical Problem Areas, Fairbanks, AK. August 8, 2001.
Mulawa, P.A., S.H. Cadle, K. Knapp, R. Zweidinger, R. Snow, R. Lucas, and J. Goldbach. 1997. Effect of ambient temperature and E-10 fuel on primary exhaust particulate matter emissions from light-duty vehicles. Environ. Sci. Technol. 31(5):1302-1307.
NRC (National Research Council). 1998. Research Priorities for Airborne Particulate Matter I: Immediate Priorities and a Long-Range Research Portfolio. Washington, DC: National Academy Press.
NRC (National Research Council). 2000. Modeling Mobile-Source Emissions. Washington, DC: National Academy Press.
NRC (National Research Council). 2001. Evaluating Vehicle Emissions Inspection and Maintenance Programs. Washington, DC: National Academy Press.
NSTC (National Science and Technology Council). 1997. Interagency Assessment of Oxygenated Fuels. National Science and Technology Council, Committee on

Environment and Natural Resources, Office of Science and Technology Policy, Executive Office of the President of the United States.

Pielke, R.A. 2002. Mesoscale Meteorological Modeling. 2nd Ed. San Diego, CA: Academic Press.

Pielke, R.A., M. Garstang, C. Lindsey, and J. Gusdorf. 1987. Use of a synoptic classification scheme to define seasons. Theor. Appl. Climatol. 38:57-68.

Pielke, R.A., R.A. Stocker, R.W. Arritt, and R.T. McNider. 1991. A procedure to estimate worst-case air quality in complex terrain. Environ. Int. 17(6):559-574.

Poloniecki, J.D., R.W. Atkinson, A.P. de Leon, and H.R. Anderson. 1997. Daily time series for cardiovascular hospital admissions and previous day's air pollution in London, UK. Occup. Environ. Med. 54(8):535-540.

Prescott, G.J., G.R. Cohen, R.A. Elton, F.G. Fowkes, and R.M. Agius. 1998. Urban air pollution and cardiopulmonary ill health: A 14.5 year time series study. Occup. Environ. Med. 55(10):697-704.

Ragazzi, R., and K. Nelson. 1999. The Impact of a 10% Ethanol Blended Fuel on the Exhaust Emissions of Tier 0 and Tier 1 Light Duty Gasoline Vehicles at 35°F. Colorado Department of Public Health and Environment, Denver, CO. March 26, 1999. [Online]. Available: http://www.cdphe.state.co.us/ap/down/oxyfuelstudy.PDF [January 30, 2002].

Raub, J.A., M. Mathieu-Nolf, N.B. Hampson, and S.R. Thom. 2000. Carbon monoxide poisoning—a public health perspective. Toxicology 145(1):1-14.

Ritz, B., and F. Yu. 1999. The effect of ambient carbon monoxide on low birth weight among children born in southern California between 1989 and 1993. Environ. Health Perspect. 107(1):17-25.

Ritz, B., F. Yu, G. Chapa, and S. Fruin. 2000. Effect of air pollution on preterm birth among children born in Southern California between 1989 and 1993. Epidemiology 11(5):502-511.

Ritz, B., F. Yu, S. Fruin, G. Chapa, G.M. Shaw, and J.A. Harris. 2002. Ambient air pollution and risk of birth defects in Southern California. Am. J. Epidemiol. 155(1):17-25.

SAE (Society of Automotive Engineers). 1992. Auto/Oil Air Quality Improvement Research Program. SAE SP-920. Warrendale, PA: Society of Automotive Engineers.

Schwartz, J. 1999. Air pollution and hospital admissions for heart disease in eight U.S. counties. Epidemiology 10(1):17-22.

Shair, F.H., and K.L. Heitner. 1974. Theoretical model for relating indoor pollutant concentrations to those outside. Environ. Sci. Technol. 8(5):444-451.

Shephard, R.J. 1983. Carbon Monoxide, Silent Killer. Springfield, IL: Charles C. Thomas.

Sheppard, L., D. Levy, G. Norris, T.V. Larson, and J.Q. Koenig. 1999. Effect of ambient air pollution on nonelderly asthma hospital admissions in Seattle, Washington, 1987-1994. Epidemiology 10(1):23-30.

Sheps, D.S., M.C. Herbst, A.L. Hinderliter, K.F. Adams, L.G. Ekelund, J.J. O'Neil, G.M. Goldstein, P.A. Bromberg, J.L. Dalton, M.N. Ballenger, S.M. Davis, and G.G. Koch. 1990. Production of arrhythmias by elevated carboxyhemoglobin in patients with coronary artery disease. Ann. Intern. Med. 113(5):343-351.

Sierra Research. 1996. Low-Temperature CO Emission Test Results. Report No. SR96-06-04. Prepared for the Alaska Department of Environmental Conservation, by Sierra Research, Inc., Sacramento, CA.

Sierra Research. 1997. Development of a Proposed Procedure for Determining the Equivalency of Alternative Inspection and Maintenance Programs. Report No. SR97-11-02. Prepared for the U.S. Environmental Protection Agency, Regional and State Programs Division, by Sierra Research, Inc., Sacramento, CA. [Online]. Available: http://www.epa.gov/ OMS/regs/im/imreadme.htm [May 30, 2001].

Sierra Research. 1999. Cold Start, Plug-in and Mid-Trip Idling CO Emissions from Light-Duty Gasoline-Powered Vehicles in Alaska. Report No. SR99-06-01. Prepared for the Alaska Department of Environmental Conservation, by Sierra Research, Inc., Sacramento, CA.

Sierra Research. 2000. Analysis of Alaska Vehicle CO Emissions Study Data. Report No. SR00-02-01. Prepared for the Municipality of Anchorage, by Sierra Research, Inc., Sacramento, CA.

Stedman, D.H., G.A. Bishop, P. Aldrete, and R.S. Slott. 1997. On-road evaluation of an automobile emission test program. Environ. Sci. Technol. 31:927-931.

Stocker, R.A., R.A. Pielke, A.J. Verdon, and J.T. Snow. 1990. Characteristics of plume releases as depicted by balloon launchings and model simulations. J. Appl. Meteorol. 29(1):53-62.

Truex, T.J. 1999. Interaction of Sulfur with Automotive Catalysts and the impact on Vehicle Emissions—A Review. SAE 1999-01-1543. Warrendale, PA: Society of Automotive Engineers.

U.S. Census Bureau. 2000. U.S. Census 2000. Summary File 2 (SF2). U.S. Census Bureau, U.S. Department of Commerce. [Online]. Available: http://www.census.gov [January 30, 2002].

U.S. Department of Commerce. 2000. Falling Through the Net: Toward Digital Inclusion: A Report on Americans' Access to Technology Tools. Washington, DC: U.S. Dept. of Commerce, Economic and Statistics Administration, National Telecommunications and Information Administration. October 2000.

Wang, J.X.L., and J.K. Angell. 1999. Air Stagnation Climatology for the United States (1948-1998). NOAA/Air Resources Laboratory Atlas No. 1. Silver Spring, MD: U.S. Dept. of Commerce, National Oceanic and Atmospheric Administration, Office of Oceanic and Atmospheric Research, Air Resources Laboratory. 73 pp.

Wayson, R.L. 1999. Dispersion Modeling at Intersections: An Overview. Presented at Microscale Air Quality Impact Assessment for Transportation Projects, Transportation Research Board, Committee on Transportation and Air Quality (A1F03), January 10, 1999.

Wenzel, T. 1999. Evaluation of Arizona's Enhanced I/M Program. Presentation at the 9th On-Road Vehicle Emissions Workshop, San Diego, CA, April 21, 1999. [Online]. Available: http://enduse.lbl.gov/projects/vehicles/Evaluation.html [May 30, 2001].

Glossary

Air-to-fuel ratio—The ratio, by weight, of air to gasoline entering the intake in a gasoline engine. The ideal (stoichiometric) ratio for complete combustion is approximately 14.7 parts of air to 1 part of fuel, depending on the composition of the specific fuel.

Air-quality model—A computer-based mathematical model used to predict air quality based on emissions and the effects of the transport, dispersion, and transformation of compounds emitted into the air.

Air toxics—Air toxics is a generic term referring to a host of carcinogens, neurotoxins, allergens, and other harmful atmospheric pollutants. The Clean Air Act Amendments of 1990 listed 189 of these air toxics as hazardous air pollutants (HAPs) for future regulation.

Alaska Department of Environmental Conservation (ADEC)—The department that deals with clean air, land, and water issues in the state of Alaska.

Albedo—The fraction of incoming sunlight that is reflected from the surface of the earth. Albedo is higher over whiter surfaces, such as snow, and lower over darker surfaces, such as oceans and forests.

Ambient air—The air outside of structures. Often used interchangeably with "outdoor air."

Box model—A model that simulates how pollutant concentrations vary over time within a designated volume of air. The effects of emissions, exchange with the surrounding atmosphere, and chemical reactions are considered. Concentrations are assumed to be well-mixed within the box.

Glossary 123

Carbon monoxide (CO)—A colorless, odorless, tasteless, and toxic gas resulting from the incomplete combustion of fuels containing carbon.

Carboxyhemoglobin—Molecule formed when CO reacts with hemoglobin, an intracellular protein that transports oxygen in the blood. The presence of carboxyhemoglobin increases the hemoglobin's affinity for oxygen, thereby reducing the transport of oxygen from the blood to the body's tissues.

Clean Air Act (CAA)—The original Clean Air Act was passed in 1963, but our national air pollution control program is actually based on the 1970 version of the law. The 1990 Clean Air Act Amendments (CAAA90) are the most recent and far-reaching revisions of the 1970 law.

Closed-loop fuel control—A fuel-metering system that uses feedback for more effective emissions control. The air-to-fuel ratio of a contemporary vehicle is "closed-loop," using a sensor in the exhaust to evaluate the mixture exiting the engine and adjusting the air-to-fuel ratio through the use of an onboard computer to optimize emissions performance.

Cold-start emissions—Tailpipe emissions that occur before a vehicle is fully warmed up. Vehicle emissions are higher during the first few minutes of operation because the engine and the catalytic converter must come to operating temperature before they can become effective.

Criteria air pollutants—A group of six common air pollutants (carbon monoxide, lead, nitrogen dioxide, ozone, particulate matter, and sulfur dioxide) regulated by the federal government since the passage of the Clean Air Act in 1970 on the basis of information on the health and/or environmental effects of each pollutant.

Cut point—For each pollutant, the emissions level above which a car is considered to have failed the emissions test for that pollutant.

Data link connector—The connector where the scan tool interfaces with a vehicle's *onboard diagnostic system.* Also know as the diagnostic link connector.

Diagnostic trouble codes—Codes stored in the engine's computer that identify emissions-control systems and components that are malfunctioning; they can be retrieved using a scan tool.

Dynamometer—A treadmill-like machine that allows cars to be tested under the loads typical of onroad driving.

Emissions budget—Allowable emissions levels identified as part of a *state implementation plan (SIP)* for pollutants emitted from mobile, industrial, stationary, and area sources. These emissions levels are used for meeting emission reduction milestones, attainment, or maintenance demonstrations.

Emissions factor—The predicted ratio of the amount of pollution produced to the amount of raw material processed or burned or of the amount of pollution produced to the activity level. By using the emission factor of a pollutant and data regarding quantities of materials used by a given source or the activity level of a given source, it is possible to compute emissions for the source. In the case of mobile source emissions, estimated emissions are the product of an emissions factor in mass of pollutant per unit distance (e.g., grams per mile) and an activity estimate in distance (e.g., average miles traveled). In the case of stationary source emissions, estimated emissions are the product of an emissions factor in mass of pollutant per unit energy (e.g., pounds per million Btu) and the amount of energy consumed.

Emissions inventory—Estimates of the amount of pollutants emitted into the atmosphere from major mobile, stationary, area-wide, and natural source categories over a specific period of time, such as a day or a year.

Empirical models—Modeling which attempts to discern a statistically significant relationship between an outcome variable (e.g., ambient CO concentration) and various predictor variables (e.g., windspeed, temperature, and inversion strength).

EPA—See *U.S. Environmental Protection Agency.*

Exceedance—An air pollution event in which the ambient concentration of a pollutant exceeds one of the National Ambient Air Quality Standards (NAAQS).

Evaporative emissions—Hydrocarbon emissions that do not come from the tailpipe of a car. Evaporative emissions can come from evaporation, permeation, seepage, and leaks in a car's fueling system. Often used interchangeably with *nontailpipe emissions.*

Federal Test Procedure (FTP)—A certification test for measuring the tailpipe and evaporative emissions from new vehicles over the Urban Dynamometer Driving Schedule, which attempts to simulate an urban driving cycle.

Gaussian dispersion models—A micro-scale model that simulates the dispersion of pollution from a source such as an intersection or a factory. The dispersion is assumed to be Gaussian in nature, and the ambient concentrations are assumed to be proportional to emissions.

Geostrophic winds—Large-scale winds in the atmosphere controlled by pressure differences. Geostrophic winds are not much influenced by the surface of the earth.

Heavy-duty vehicles (HDV)—Any motor vehicle rated at more than 8,500 lb gross vehicle weight (GVWR) or that has a vehicle curb weight of more than 6,000 lb or that has a basic vehicle frontal area in excess of 45 ft^2. This excludes vehicles that will be classified as medium-duty passenger vehicles for the purposes of the Tier 2 emissions standards.

Heavy-duty diesel vehicles (HDDV)—An HDV that uses diesel fuel.

Hydrocarbons (HC)—Organic compounds containing various combinations of hydrogen and carbon.

Ice fog—A meteorological phenomenon that occurs in Fairbanks, Alaska, on very cold days (colder than -20°F). Tiny ice particles are formed in the air when water vapor released from cars, humans, or power plants is frozen very quickly. If an elevated inversion is present, the ice fog can be trapped near the surface.

Inspection and maintenance (I/M) programs—State emissions testings programs that attempt to identify vehicles with higher than allowable emissions and ensure that such vehicles are repaired or removed from the fleet.

Inversion—See *Temperature inversion*.

IM240—The name for the emissions test used in some I/M programs, including those in Arizona and Colorado. The IM240 is a *transient, loaded-mode emissions test*. "Loaded-mode" refers to the fact that the test is run on a treadmill-like device called a *dynamometer*, which simulates driving with the engine in gear. "Transient" refers to the fact that the car drives under a load that varies from second to second during the test. The "240" in IM240 indicates that the test lasts for 240 seconds. The IM240 is intended by EPA to be a shortened version of part of the FTP and to correlate well with the FTP.

Lapse rate—The rate at which temperature in the atmosphere changes with altitude. The average lapse rate is about -6.5°C/km. Under inversion conditions, the lapse rate can be positive.

Light-duty vehicle (LDV)—A passenger car or passenger car derivative capable of seating 12 or fewer passengers. All vehicles and trucks under 8,500 lb GVWR are included (this limit previously was 6,000 lb). Small pickup trucks, vans, and sport utility vehicles may be included.

Loaded-mode emissions test—An emissions test performed with the engine in gear.

Malfunction indicator light (MIL)—The instrument panel light used by the *onboard diagnostic system* to notify the vehicle operator of an emissions related fault. The MIL is also known as the "service engine soon" or "check engine" lamp.

Medium-duty passenger vehicle (MDPV)—A new class of vehicles introduced with the Tier 2 emissions standards that includes sport utility vehicles and passenger vans rated at between 8,500 and 10,000 lb GVWR.

Mixing layer—A layer of air near the surface in which gaseous pollutants are well-mixed.

Model year—Vehicles are certified for sale, marketed, and later registered as a certain model year, indicating the year a vehicle was produced and offered for sale. Model years typically begin in September or October of the prior year and run for roughly 12 months. In the last decade, certain vehicles have been introduced as "pull-ahead" vehicles, appearing as early as January of the preceding year.

National Ambient Air Quality Standards (NAAQS)—Standards set by EPA for the maximum levels of criteria air pollutants that can exist in the outdoor air without unacceptable effects on human health or the public welfare.

Nonattainment area—A geographic area in which the concentration of a criteria air pollutant has exceeded the concentration allowed by the federal standards at some recent time. A single geographic area may have acceptable concentrations of one criteria air pollutant but unacceptable concentrations of one or more other pollutants; thus, an area can be both an attainment area and a nonattainment area at the same time.

Nitrogen oxides (NO_x)—A general term referring to nitric oxide (NO) and nitrogen dioxide (NO_2). Nitrogen oxides are formed when air is raised to high temperatures, such as during combustion or lightning, and are major contributors to smog formation and acid deposition.

Numerical predictive model—A meso-scale or large-scale model used to predict chemical concentrations in the atmosphere based on observed meteorological variables, emissions, and chemistry. Numerical predictive models represent the atmosphere as a three-dimensional grid of air parcels. Chemical transformations take place within the air parcels, and air is transported between them.

O_2 sensor—See *Oxygen sensor.*

OBDI—See *Onboard diagnostics generation I.*

OBDII—See *Onboard diagnostics generation II.*

Onboard diagnostic system—A device incorporated into the computers of new motor vehicles to monitor the performance of the emission controls. The computer triggers a dashboard indicator light, referred to as a malfunction indicator light, when the controls malfunction, alerting the driver to seek maintenance for the vehicle. The system also communicates its findings to repair technicians by means of diagnostics trouble codes, which can be downloaded from the vehicle's computer. Onboard diagnostic systems do not actually measure emissions.

Glossary 127

Onboard diagnostics generation I (OBDI)—An onboard automotive diagnostic system required by the California Air Resources Board since 1988. It uses the microprocessor and sensors to monitor and control various engine system functions.

Onboard diagnostics generation II (OBDII)—OBDII expands upon OBDI to include emissions-system and sensor-deterioration monitoring.

Open-loop fuel control—A system in which the air-fuel mixture is preset by design and contains no feedback correction signal to optimize fuel metering for emissions control. (See also *Closed-loop fuel control*.)

Oxygen (O_2) sensor—A sensor placed in the exhaust that measures exhaust oxygen content. Typically, there are oxygen sensors before and after the catalytic converter.

Oxygenated fuel (oxyfuel)—Gasoline containing oxygenates, typically methyl *tertiary*-butyl ether (MTBE) or ethanol, intended to reduce production of carbon monoxide, a criteria air pollutant. In some parts of the country, carbon monoxide emissions from cars make a major contribution to pollution. In some of those areas, gasoline refiners must market oxygenated fuels, which typically contain 2-3% oxygen by weight.

Oxygenates—Compounds containing oxygen (alcohols and ethers) that are added to gasoline to increase its oxygen content. Methyl *tertiary*-butyl ether (MTBE) and ethanol are the most common oxygenates currently used, although there are a number of others.

Ozone (O_3)—A reactive gas whose molecules contain three oxygen atoms. It is a product of photochemical processes involving sunlight and ozone precursors, such as hydrocarbons and oxides of nitrogen. Ozone exists in the upper atmosphere (stratospheric ozone), where it helps shied the earth from excessive ultraviolet rays, as well as in the lower atmosphere near the earth's surface (tropospheric ozone). Tropospheric ozone causes plant damage and adverse health effects and is a criteria air pollutant. Tropospheric ozone is a major component of smog.

Particulate matter (PM)—Any material, except uncombined water, that exists in solid or liquid states in the atmosphere. The size of particulate matter can vary from coarse, wind-blown dust particles to fine particles directly emitted as combustion products or formed through secondary reactions in the atmosphere.

Photochemical reaction—A term referring to a chemical reaction brought about by the light of the sun. The formation of ozone from nitrogen oxides and hydrocarbons in the presence of sunlight involves photochemical reactions.

Plug-in—An electric device used to heat the engine in order to facilitate engine starting and reduce the time for emissions-control devices to be activated.

$PM_{2.5}$—A subset of particulate matter that includes those tiny particles with an aerodynamic diameter less than or equal to a nominal 2.5 micrometers. This fraction of particulate matter penetrates most deeply into the lungs and causes the majority of visibility reduction.

PM_{10}—A major air pollutant consisting of small particles with an aerodynamic diameter less than or equal to a nominal ten micrometers (about one-seventh the diameter of a single human hair). Their small size allows them to make their way to the air sacs deep within the lungs where they may be deposited and result in adverse health effects. PM_{10} also causes visibility reduction.

Primary standard—A NAAQS for criteria air pollutants based on health effects.

Process numerical model—A meso-scale or large-scale model used to analyze atmospheric processes and their effect on air quality, typically for research purposes. Like *numerical predictive models*, process numerical models represent the atmosphere as a three-dimensional grid of air parcels. Process numerical models better resolve the coupling between meteorology and chemistry than do numerical predictive models and are not necessarily constrained by observations.

Reformulated gasoline (RFG)—Specifically formulated fuel blended such that, on average, the exhaust and evaporative emissions of *volatile organic carbons* (VOCs) and hazardous air pollutants (chiefly benzene, 1,3-butadiene, *polycyclic aromatic hydrocarbons*, formaldehyde, and acetaldehyde) resulting from RFG use in motor vehicles might be significantly and consistently lower than such emissions resulting from use of conventional gasolines. The 1990 Clean Air Act Amendments require sale of RFG in the nine areas with the most severe ozone pollution problems. RFG contains, on average, a minimum of 2.0 weight percent oxygen.

Remote sensing—A method for measuring pollution levels in a vehicle's exhaust while the vehicle is traveling down the road. Remote-sensing systems employ infrared absorption to measure VOC and carbon monoxide emissions relative to carbon dioxide. These systems typically operate by continuously projecting a beam of infrared radiation across a roadway and making measurements on the exhaust plume after a vehicle passes through the beam.

Scan tool—A hand-held computer that is plugged into a vehicle's *data link connector* allowing a technician to read diagnostic trouble codes, readiness status, and other information collected by the *onboard diagnostic system*.

Secondary particle—Particulate matter that is formed in the atmosphere and is generally composed of species such as ammonia or the products of atmospheric chemical reactions such as nitrates, sulfates, and organic material. Secondary particles are distinguished from primary particles, which are emitted directly into the atmosphere.

Secondary standard—A NAAQS for criteria air pollutants based on environmental effects such as damage to property, plants, or visibility.

Supplemental Federal Test Procedure (SFTP)—The SFTP is a certification test for measuring the tailpipe and evaporative emissions from new vehicles that includes two driving cycles not represented in the FTP. The SFTP includes a test cycle simulating high speed and high acceleration driving (US06 cycle) and a test cycle that evaluates the effects of air conditioner operation (SC03 cycle).

State implementation plan (SIP)—A detailed description of the programs a state will use to carry out its responsibilities under the Clean Air Act for complying with the NAAQS. SIPs are a collection of the programs used by a state to reduce air pollution. The Clean Air Act requires that EPA approve each SIP. The public is given opportunities to participate in the review and approval of SIPs.

Statistical roll-back model—A model that estimates the emissions reductions needed for a desired improvement in air quality. The emissions reductions are assumed to be linearly related to ambient concentrations of pollutants.

Steady-state emissions test—An emissions test performed under one stable operating condition, such as when a vehicle is tested at idle or under a constant engine load.

Subsidence—Slow descent of air due to radiative cooling.

Synoptic—Used to describe meteorological processes that occur over regional spatial extents and over several days.

Tampering—The malfunctioning of one or more emissions-control devices due to either deliberate disablement or mechanical failure.

Temperature inversion—An atmospheric condition in which temperature in the lower part of the atmosphere increases with altitude, rather than decreasing with altitude as is more typical. Inversion conditions can trap pollution near the surface because warmer, less-dense air is resting above colder, more-dense air.

Three-way catalytic converter—A catalytic converter designed to oxidize CO and VOCs and reduce NO_x emissions from gasoline-fueled vehicles.

Transient emissions test—An emissions test performed under a load that varies from moment to moment during the test.

Transportation control measure (TCM)—Any control measure to reduce vehicle trips, vehicle use, vehicle miles traveled, vehicle idling, or traffic congestion for the purpose of reducing motor-vehicle emissions. TCMs can include encouraging the use of carpools and mass transit.

Transportation-demand management (TDM) strategies—Strategies which use regulatory mandates, economic incentives, or education campaigns to change driver behavior. TDM strategies attempt to reduce the frequency or length of automobile trips or to shift the timing of automobile trips.

Transportation-supply improvement (TSI) strategies—TSI strategies attempt to reduce emissions by changing the physical infrastructure of the road system to improve traffic flow and reduce stop-and-go movements.

Two-way catalytic converter—A first generation catalytic converter designed to oxidize CO and VOC emissions from gasoline-fueled vehicles.

Urban airshed model (UAM)—A three-dimensional photochemical air quality grid model for calculating the concentrations of both inert and chemically reactive pollutants in the atmosphere. It simulates the physical and chemical processes that affect concentrations of pollutants.

U.S. Environmental Protection Agency (EPA)—The federal government agency that establishes regulations and oversees the enforcement of laws related to the environment.

Vehicle-miles traveled (VMT)—The number of miles driven by a fleet of vehicles over a set period of time, such as a day, month, or year.

Sources: California Air Resources Board 2002; Davis 1997; EPA 2002; FHWA 2000; Harvey and Deakin 1993; IMRC 2000.

Appendix

Biographical Information on the Committee on Carbon Monoxide Episodes in Meteorological and Topographical Problem Areas

Armistead G. Russell (*Chair*) is the Georgia Power Distinguished Professor of Environmental Engineering at the Georgia Institute of Technology. His research areas include air pollution control, aerosol dynamics, atmospheric chemistry, emissions control, air pollution control strategy design and computer modeling. Dr. Russell has served on a number of National Research Council (NRC) committees and was chair of the Committee to Review EPA's Mobile Source Emissions Factor (MOBILE) model. He received a Ph.D. in mechanical engineering from the California Institute of Technology.

Roger Atkinson is a research chemist and Distinguished Professor of Atmospheric Chemistry at the University of California, Riverside. His research areas include the kinetics and mechanisms of atmospherically important reactions of organic compounds in the gas phase. Dr. Atkinson serves on the California Air Resources Board's Reactivity Scientific Advisory Committee and its Scientific Review Panel on Air Toxics, and he has served on NRC committees including the Committee on Tropospheric Ozone Formation and Measurement and the Committee on Ozone-Forming Potential of Reformulated Gasoline. He received a Ph.D in physical chemistry from the University of Cambridge.

Sue Ann Bowling is a retired professor at the University of Alaska, Fairbanks. Her research interests included air pollution meteorology, polar meteorology, radiative transfer, paleoclimatology, and climatic change. Dr. Bowling received a Ph.D. from the University of Alaska, Fairbanks.

Steven D. Colome is deputy director of the Southern California Particle Center and Supersite and an adjunct professor in environmental health at the University of California, Los Angeles, School of Public Health. His research interests include human exposure assessment, environmental epidemiology, indoor air quality, regional exposure modeling, and health effects assessment. Dr. Colome previously served on the NRC Committee on Toxicological and Performance Aspects of Oxygenated Motor Vehicle Fuels. He was a reviewer of the EPA document Air Quality Criteria for Carbon Monoxide. He received a Sc.D. in environmental health sciences from Harvard University, School of Public Health.

Naihua Duan is professor in residence in the Department of Psychiatry and Biobehavioral Sciences and the Department of Biostatistics at the University of California, Los Angeles. Previously, he was a corporate chair and Senior RAND Fellow in statistics at RAND. His research interests include nonparametric and semiparametric regression methods, sample design, hierarchical models, and environmental exposure assessment, including exposure to carbon monoxide. He served as a member of the NRC Committee on Advances in Assessing Human Exposure to Airborne Pollutants. Dr. Duan received a Ph.D. in statistics from Stanford University.

Gerald Gallagher is president of J. Gallagher and Associates. Previously, he served as manager of the Mobile Sources Program for the Air Pollution Control Division of the Colorado Department of Public Health and Environment. His responsibilities included the development and implementation of air quality management plans for controlling carbon monoxide. He was also responsible for the operation of a metro-wide inspection and maintenance program, consisting of about 1.8 million inspections per year for gasoline- and diesel-fueled vehicles. Dr. Gallagher is a member of the NRC Committee on Vehicle Emissions Inspection and Maintenance Programs. He received a Ph.D. in intergovernmental relations and environmental management from the University of Colorado.

Appendix 133

Randall L. Guensler is an associate professor in the School of Civil and Environmental Engineering at Georgia Institute of Technology. His research interests include the relationships between land use, infrastructure, travel behavior, and vehicle emission rates; transportation and air quality planning and modeling theory and practice; and emission control-strategy effectiveness. Dr. Guensler is currently chairman of the Transportation Research Board's Committee on Transportation and Air Quality. He received a Ph.D. in civil and environmental engineering from the University of California, Davis.

Susan L. Handy is an associate professor in the Community and Regional Planning Program at the University of Texas. Her research focuses on the relationship between transportation systems and land use patterns, and the effects of telecommunication technologies on patterns of development and travel behavior. Dr. Handy is chair of the Transportation Research Board Committee on Telecommunications and Travel Behavior and also serves on the Committee on Transportation and Land Development. She received a Ph.D. in city and regional planning from the University of California, Berkeley.

Simone Hochgreb is managing engineer in the Thermal Sciences Practice at Exponent. Her research focuses on fundamental and applied problems in combustion and chemical kinetics, particularly applications to transportation, internal-combustion engines, and pollutant-emission formation. Dr. Hochgreb served as a member of the NRC Committee on Toxicological and Performance Aspects of Oxygenated Motor Vehicle Fuels and the NRC Review Panel for the Partnership for a New Generation of Vehicles. She received a Ph.D. in mechanical and aerospace engineering from Princeton University.

Sandra N. Mohr is an assistant professor at the University of Colorado Health Sciences Center and is the residency director for occupational and environmental medicine there and at the National Jewish Medical and Research Center. Dr. Mohr's research focuses on the health effects of air pollutants. She has been a lead researcher in the health effects of methyl *tertiary*-butyl ether (MTBE), a gasoline additive, and has served on the NRC Committee on Toxicological and Performance Aspects of Oxygenated Motor Vehicle Fuels. She received an M.D. from the University of Kansas School of Medicine and an M.P.H. degree from Yale University.

Roger A. Pielke Sr. is a professor in the Department of Atmospheric Science at Colorado State University. He is also state climatologist for Colorado. His research areas include the study of global, regional, and local weather and climate phenomena through the use of sophisticated mathematical simulation models, air pollution meteorology, and mesoscale meteorology. Dr. Pielke received a Ph.D. in meteorology from Pennsylvania State University.

Karl J. Springer is retired vice president for automotive products and emissions research at Southwest Research Institute. His research focused on the measurement and control of air pollution emissions from on-road and off-road vehicles and equipment powered by internal-combustion engines. Mr. Springer is a member of the National Academy of Engineering. He received a BSME from Texas A & M and an M.S. in physics from Trinity University.

Roger Wayson is an associate professor of civil and environmental engineering at the University of Central Florida, where he conducts research in the microscale modeling of carbon monoxide ambient concentrations that result from mobile sources and airport operations. Dr. Wayson obtained his B.S. and M.S. in environmental engineering from the University of Texas, Austin, and his Ph.D. in civil engineering from Vanderbilt University.